Topics in Mining, Metallurgy and Materials Engineering

Series Editor

Carlos P. Bergmann, Federal University of Rio Grande do Sul, Porto Alegre, Rio Grande do Sul, Brazil

"Topics in Mining, Metallurgy and Materials Engineering" welcomes manuscripts in these three main focus areas: Extractive Metallurgy/Mineral Technology; Manufacturing Processes, and Materials Science and Technology. Manuscripts should present scientific solutions for technological problems. The three focus areas have a vertically lined multidisciplinarity, starting from mineral assets, their extraction and processing, their transformation into materials useful for the society, and their interaction with the environment.

**** Indexed by Scopus (2020) ****

More information about this series at http://www.springer.com/series/11054

Sabrina Arcaro · Janio Venturini

Modern Ferrites in Engineering

Synthesis, Processing and Cutting-Edge Applications

Springer

Sabrina Arcaro
Universidade do Extremo Sul Catarinense
Criciúma, Santa Catarina, Brazil

Janio Venturini
Federal University of Rio Grande do Sul
Porto Alegre, Rio Grande do Sul, Brazil

ISSN 2364-3293 ISSN 2364-3307 (electronic)
Topics in Mining, Metallurgy and Materials Engineering
ISBN 978-3-030-78987-9 ISBN 978-3-030-78988-6 (eBook)
https://doi.org/10.1007/978-3-030-78988-6

This Springer imprint is published by the registered company Springer Nature Switzerland AG
The registered company address is: Gewerbestrasse 11, 6330 Cham, Switzerland

Contents

Chapter 1
A Brief History of Ferrites

Magnetic ceramic materials composed mainly of iron (III) oxide are commonly known as "ferrites." The first of these materials known to men was lodestone, a mineral form of magnetite (Fe_3O_4) that is naturally magnetized. One of the earliest reports of the unique properties of this ore comes from Thales of Miletus. In the sixth century BC, the philosopher described the attraction of lodestone to iron and other lodestones. Ores from the region around the Greek town of Magnesia, in Asia Minor, are usually considered as the source of the term "magnet." The attraction phenomenon was also known to the Chinese by the late Warring States period (fourth century BC), as referenced in the *Guiguzi*. By the first centuries of the current era, Chinese divinators reportedly used compass-like devices made of lodestone for their esoteric rituals.

William Gilbert gave us one of the earliest modern scientific descriptions of the properties of lodestones in his "*De Magnete, Magneticisque Corporibus, et de Magno Magnete Tellure*," published in 1600. In this treatise, Gilbert proposed that the planet behaved like a giant magnet instead of a single region in the poles. He studied these properties via a scale model of the planet (*terrella*, Latin for "small Earth") made of lodestone. The scientist also claimed that magnetism was the force holding the moon in orbit around Earth. Albeit erroneous, this was one of the first attempts at a scientific explanation for the motion of heavenly bodies as a counterpoint to Aristotelian astronomy.

The work of Takei and Kato kickstarted the utilization of ferrites in electrical appliances. In 1930, the Tokyo Institute of Technology scientists discovered that oxides containing iron and divalent cations had distinctive magnetic properties. Their discovery of these "ferrites" was presented at the Annual Meeting of the American Electrochemical in St Louis [1]. Although initially unsure whether these materials would find a commercial application, their work led to the creation of the TDK Corporation in 1935. Ferrite cores were produced and massively applied for the first time in 1937, mostly in wireless communication devices. By the end of World War II, the company had already produced 5 million ferrite cores, primarily to aid the Japanese war effort.

S. Arcaro and J. Venturini, *Modern Ferrites in Engineering*, Topics in Mining, Metallurgy and Materials Engineering, https://doi.org/10.1007/978-3-030-78988-6_1

In 1948, Neel gave his theoretical description of ferrimagnetism. His work successfully explained the magnetic properties of the zinc ferrites developed by Takei and Kato. The French scientist proposed a sublattice model in which the spins in each crystalline subunit interact with one another, leading to the intriguing magnetic behavior of ferrites. These developments created the conditions for the massive industrial application of ferrite-based electronic devices.

Studies conducted at the Philips laboratories in the Netherlands in the 1940s and 1950s also contributed decisively toward the advancement of ferrites. Snoeck and Verwey are some of the famous researchers from this era that helped create materials with high electrical resistivity and low relaxation losses. Their discoveries led to the quick development and adoption of ferrite cores. The laboratory also created several of the most utilized hexaferrites. In 1950, they created barium ferrite ($BaFe_{12}O_{19}$), a low-cost hexaferrite with elevated coercivity. A decade later, Philips developed strontium hexaferrite, a similar material with improved properties. Both hexaferrites now dominate the worldwide market for these materials.

In the year since, ferrites have become ubiquitous in most electronic devices. Antennas, ferrite cores, and EMI suppressors are examples of applied ferrites that one will undoubtedly interact with in our modern daily life. Nevertheless, the field of application of these materials has been broadening ever since their embryonic stages. Nowadays, the application of ferrites encompasses a wide range of commercial purposes, from the medical field to quantum computing. Given their low cost and unusual properties, it is no surprise that these materials have attracted the industry's interest.

Researchers worldwide are striving to find novel properties and applications for ferrites. Figure 1.1 shows the number of publications indexed in the ScienceDirect portal containing the terms "spinel" and "ferrite." As can be seen in the graph, there is growing interest from researchers worldwide in the development of these extraordinary materials. The number of scientific articles on these ceramics has grown almost ten times in the period between 2000 and 2020.

A bibliometric network of the publications from the same period (Fig. 1.2) gives us insight into the varied research areas focusing on ferrites. This network was created from co-citations, defined as the frequency with which two documents are cited simultaneously by other documents. The manuscripts were extracted from the Scopus database, and the minimum number of co-citations was 10. In general, much effort is still directed at improving their electrical and magnetic properties. However, ferrites are also considered promising materials for application in very disparate fields, such as in biomedicine and lithium-ion batteries, as seen in the network. Undoubtedly, many works focus on studying the structures of ferrites since the structure strongly influences the properties. The synthesis of pure ferrites and their composites is also given great attention. Another focal point for research groups worldwide is the utilization of these materials as magnetic catalysts in organic reactions, as the ease of magnetic separation allows their utilization as unsupported catalysts in industrial reactors.

The 15 countries that have published the most articles with the keywords "ferrite" and "spinel" in the last 20 years are shown in Fig. 1.3. A significant share of research

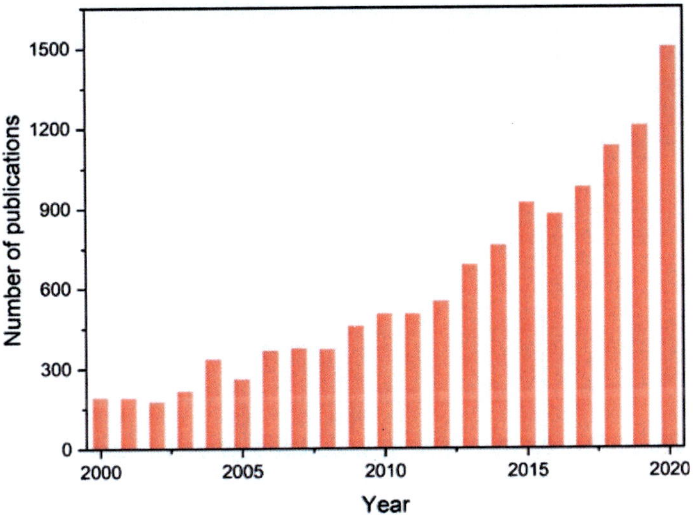

Fig. 1.1 Number of publications in the last twenty years containing the terms "spinel" and "ferrite" indexed in the ScienceDirect database

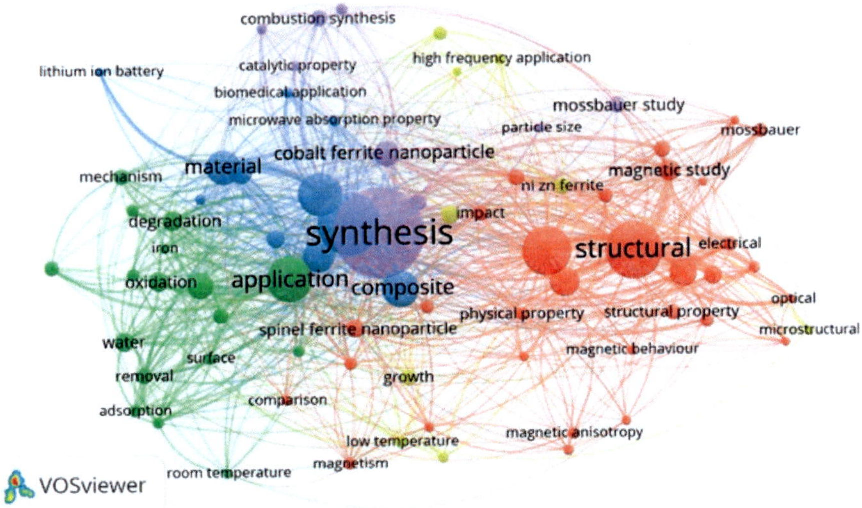

Fig. 1.2 Bibliometric network of publications containing the term "spinel ferrites" in the period between 2000 and 2020. Larger circles indicate a larger prevalence of the terms in the title or abstract of the article. Produced with the aid of the VOSviewer software

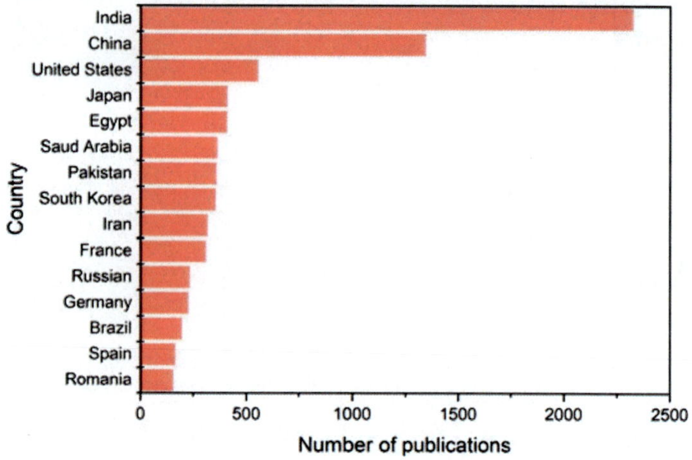

Fig. 1.3 Top 15 countries with the most publications containing the terms "spinel" and "ferrite" indexed in the ScienceDirect database in the last twenty years

in the field of ferrites is carried out in India and China, followed closely by the United States. Other countries such as Japan, France, and South Korea also have prominent publications in this area.

In summary, ferrites are an essential tool in the industrial arsenal ever since they were first mass-produced in the 1930s. This book aims to discuss several of the most relevant properties of these ceramics, as well as their production methods. Selected applications of ferrites will also be addressed.

Reference

1. Kato, Y, Takei, T (1930) Studies on zinc ferrite. Its formation, composition, and chemical and magnetic properties. J Electrochem Soc 57:297–312. https://doi.org/10.1149/1.3492161

Chapter 2
Structure of Ferrites

This chapter discusses the main families of ferrites, spinel, and hexaferrites, as well as some of the other possible crystal structures displayed by these materials. The structure of spinel ferrites and the magnetic consequences of this arrangement are thoroughly covered, along with a discussion on the methods utilized in the modification of the inversion degrees of these ferrimagnets. The structural units of hexaferrites and their main subclassifications are presented. Finally, the crystal structure of perovskite and garnet ferrites is discussed, along with the presentation of their most illustrious members.

2.1 Introduction

Ferrites are a group of materials that are composed of iron (III) oxide (Fe_2O_3) and another transition metal oxide. Their composition can thus be said to be $MeO_x.Fe_2O_3$, where Me is a d-block metal. In general, the material is said to be a ferrite based not only on its composition but also by their display of pronounced magnetic properties.

The family of the ferrites encompasses a wide range of ceramic compositions. Nevertheless, the majority of these materials can be categorized into one of the main ferrites subsets. The first one contains those materials displaying cubic symmetry, the *spinel ferrites*. Many of the remaining ferrites exhibit hexagonal symmetry and are therefore aptly named *hexaferrites.* A few other crystal arrangements are also widely considered as ferrites and, as such, will likewise be discussed in this chapter.

2.2 Spinel Ferrites

The spinel ferrites are the most readily recognizable ferrites. The most familiar member of this group, magnetite (Fe_3O_4), is not only the archetypal ferrite but indeed

S. Arcaro and J. Venturini, *Modern Ferrites in Engineering*, Topics in Mining, Metallurgy and Materials Engineering, https://doi.org/10.1007/978-3-030-78988-6_2

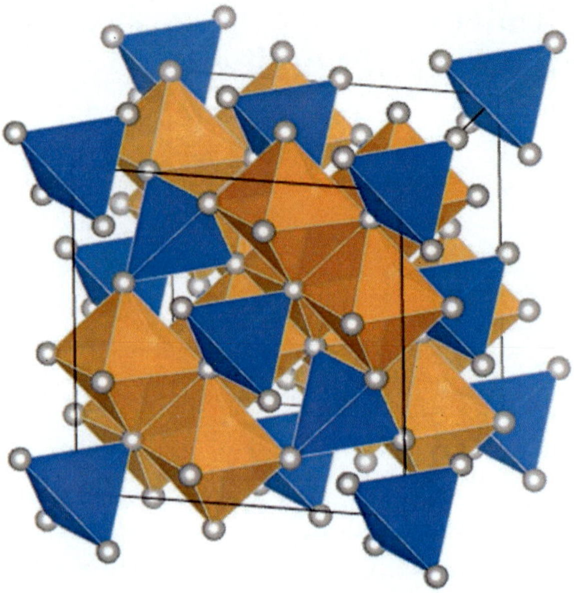

Fig. 2.1 Crystalline structure of spinel ferrites

the earliest material recognized to be magnetic, giving birth to the term "magnetism" itself.

Spinel ferrites can be described by the formula MFe_2O_4, where M is usually a divalent cation, such as cobalt or zinc, or a mixture of cations with an average charge of two (i.e., Li^+ and Fe^{3+}). These materials can display a structural configuration analogous to that of spinel ($MgAl_2O_4$), hence the name of this group. In this AB_2O_4 atomic arrangement, a face-centered cubic lattice of O^{2-} ions is partially filled by the metallic cations. More specifically, one-sixteenth of the tetrahedral positions (site A) is filled by the divalent cations, while one-eighth of the octahedral B sites is populated by trivalent metals—Fe^{3+} in the case of spinel ferrites. The crystal symmetry of this structure can be seen in Fig. 2.1.

Another very common configuration of ferrites is the inverse spinel arrangement. In this case, the tetrahedral positions are all filled by the trivalent iron atoms (Fe^{3+}), while the octahedral voids are populated by the remaining iron atoms and the divalent cation. Such configuration is widespread in some usual ferrites, such as $CoFe_2O_4$ and Fe_3O_4 in their bulk forms.

In reality, most ferrites are neither completely normal nor inverse spinels. In most cases, ferrites adopt a mixed configuration, i.e., an intermediate structure between those two extremes. Their crystalline arrangement can thus be characterized by its degree of inversion x. This parameter ranges from zero (normal spinel) to one, in the case of a wholly inverted spinel. Ferrites that are commonly considered as normal spinels tend to display values that are not precisely zero, given that slight deviations from this value incur a considerable increase in configurational entropy, thus stabilizing the structure. The same can be said of inverse spinels.

The inversion degree of spinel ferrites is of particular importance due to the ferrimagnetic nature of these materials. The major contributor toward the spontaneous spin alignment comes from AB interactions. In other words, the exchange interaction between the tetrahedral and the octahedral sites is the primary driver of the net magnetic alignment in spinel ferrites. These AB interactions are the closest to an ideal angle of 180° and have short Me-O distances, both of which work toward favoring this exchange geometry. BB exchange interactions are much weaker, and the AA situation is even worse, with an A-O-A angle of approximately 80°, thus decreasing the contribution of this exchange integral. The geometries of all possible interactions are depicted in Fig. 2.2. Néel himself explained the ferrimagnetism of most spinel ferrites in terms of the antiparallel alignment arising from the AB interaction. For the magnetism of the ferrites to be consistent throughout the crystal, the entire A (tetrahedral) sublattice arranges its spins antiparallel to those of the B (octahedral) sites. Therefore, in a perfect crystal, each of the sublattices behaves as a separate block with a self-consistent alignment. Consequently, in materials where cations with unpaired spins populate both A and B sublattices, as is the case of most spinel ferrites, the material displays ferrimagnetism. This term derives its name itself from the word "ferrite," the materials which were first observed to possess this form of magnetic alignment.

Given the prominence of the AB interaction toward the overall magnetism of spinel ferrites, the importance of the inversion degree of the structure becomes evident. In the case of only one of the positions being occupied by a magnetic cation, as is the case for zinc ferrite, no net alignment is observed. In this material, an inversion of zero is commonly observed; that is, the structure is a normal spinel. The tetrahedral A position is occupied by non-magnetic Zn^{2+}, which leaves the remaining magnetic Fe^{3+} ions in the octahedral sites. In this structure, no AB interaction is possible, and the remaining BB exchange (Fe^{3+}-Fe^{3+}) is not strong enough for the material to display any magnetism at standard conditions.

Even in ordinarily magnetic ferrites, such as cobalt ferrite ($CoFe_2O_4$), the inversion degree largely dictates the strength of the magnetism displayed by the sample.

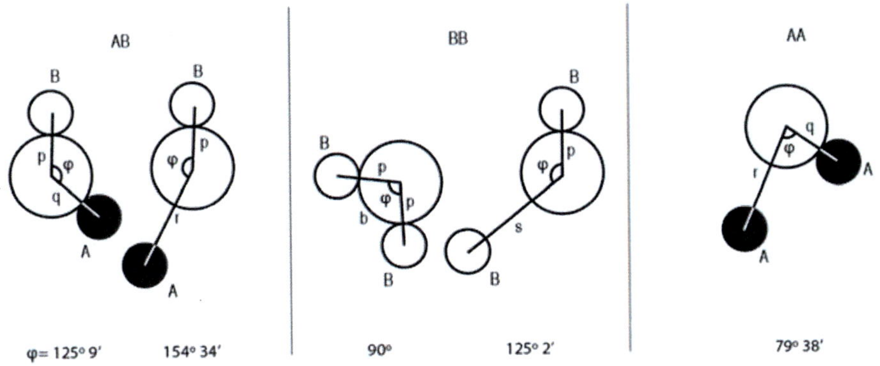

Fig. 2.2 Geometries of the possible interactions in the spinel structure

Table 2.1 Magnetic substructure in selected spinel ferrites

Ferrite	Tetrahedral lattice	Octahedral Lattice	Approximate net magnetic moment per unit formula (μ_B)
$ZnFe_2O_4$	Zn^{2+}	Fe^{3+} Fe^{3+}	0
$CoFe_2O_4$ (inverse)	Co^{2+} Fe^{3+}	Fe^{3+}	3
$CoFe_2O_4$ (normal)	Co^{2+}	Fe^{3+} Fe^{3+}	7

In a normal configuration, the much more substantial contribution of octahedral Fe^{3+} (5 unpaired spins each) is offset by the unpaired spins of Co^{2+} (3 μ_B per ion), which should result in an approximate net magnetic moment of 7 μ_B per unit formula. On the other hand, in an inverse arrangement, the magnetic contribution of the iron ions in the antiparallel octahedral and tetrahedral sublattices should cancel each other out, with the net magnetism being a function solely of the divalent component, resulting in a net magnetism of 3 μ_B per unit formula. A scheme depicting the summation of the antiparallel spins in each lattice can be seen in Table 2.1.

The preference of each cation toward each of the spinel sites was usually explained in terms of the CFSE (crystal field stabilization energy) theory. Some cations show a marked tendency toward populating either the octahedral or tetrahedral sites depending on the structure of their d-bands. If the divalent cation is much more stable in an octahedral environment, it should displace half of the B^{3+} ions, creating an inverse spinel. A similar situation would be created if the B ions have a low octahedral site stabilization energy, expelling the A^{2+} ions from the tetrahedral positions. Nevertheless, such explanations are nowadays taken with a pinch of salt; there is a plethora of examples of spinel arrangements that cannot be explained by the CFSE theory alone.

Size effects have also been utilized in explaining the site preference of different cations. Larger ions have a tendency toward occupying the octahedral site of the spinel lattice (0.414R, with R being the radius of the anion) while the smallest cation should be forced to fill the smaller tetrahedral void (0.225R). Charge effects can also account for the population of the sublattices; the trivalent ion would be better stabilized in an octahedral position, where it would be surrounded by six anions, than in the tetrahedral environment, with only four oxygen ions around it. The opposite could be said of the divalent cation, which would be forced into a tetrahedral configuration. In practice, a sum of these effects should dictate the inversion degree of the ferrites.

Nevertheless, kinetic effects show a much stronger influence than the expected thermodynamic factors. Quenching from high temperatures, for instance, is one of the strategies utilized in "freezing" different cation configurations in place. When rapidly cooled, the crystalline framework of the ferrites does not have enough time to relax toward its most thermodynamically stable state. The cations hold the positions that they would occupy at the higher temperature, therefore presenting an inversion degree that would not normally be expected under the studied conditions. For instance,

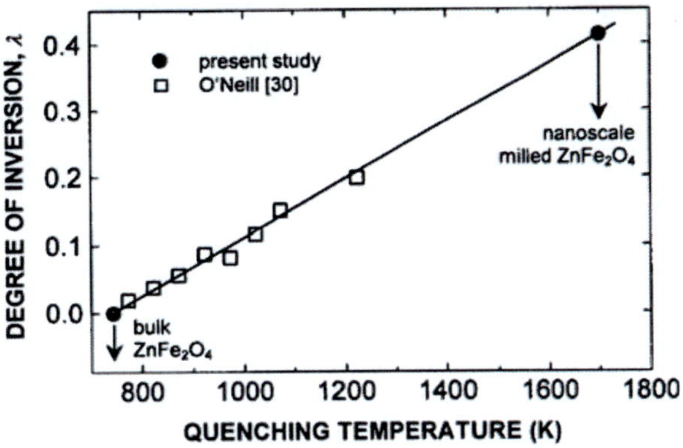

Fig. 2.3 Dependence of the inversion degree on the quenching temperature for $ZnFe_2O_4$. Data was obtained from the work of [2] (Reprinted from [3] with permission from Elsevier)

Faller and Birchenall [1] published a fascinating study on the effect of quenching on the cation distribution of magnesium and nickel ferrite and the relation of the inversion parameter with the temperature. A comparison of their results with the Boltzmann relation according to Néel can be seen in their manuscript. A relation between quenching temperature and degree of inversion for $ZnFe_2O_4$ is shown in Fig. 2.3.

Mechanical stresses have also been shown to lead to metastable inversion degrees. Sepelák and collaborators [4] have demonstrated the effect of high-energy milling on the cationic distribution inside magnesium ferrites. The mechanical stresses induced by the process led to a pronounced decrease in the inversion of the studied ferrites. The effect was comparable to that obtained by quenching from 900 °C. The thermal stability of the induced effect was also analyzed. The authors report recrystallization of the powders occurs above 600 K, above which the system relaxes toward thermodynamic equilibrium.

The synthesis method chosen in the preparation of ferrites is also widely regarded as crucial in the definition of the inversion degree. Dolcet et al. [5] have recently presented a very interesting work on different wet chemical approaches toward the production of $ZnFe_2O_4$ nanoparticles. The researchers have shown, for instance, that a combined microemulsion and hydrothermal approach favors the transfer of Zn^{2+} atoms to the octahedral sites. On the other hand, when using a microwave method, the different crystallization mechanisms lead to the production of frameworks with a stronger normal character. Furthermore, the article also shows that the inversion degree can also be tuned to a certain extent by varying the duration of the thermal step applied during the synthesis. The extraction of the inversion degrees relied on the Raman spectra of the produced materials, which are reproduced in Fig. 2.4.

Despite seeming to be a rather straightforward property, the determination of the inversion degrees of spinel ferrites poses quite a serious challenge. Several techniques are utilized in this determination, to varying degrees of success. X-ray diffractometry is often reported in the literature as one of the simplest methods for the estimation of the inversion degree of ferrites. Podwórny [6] recently published a fascinating study on the determination of the site occupancy in related spinels based on several

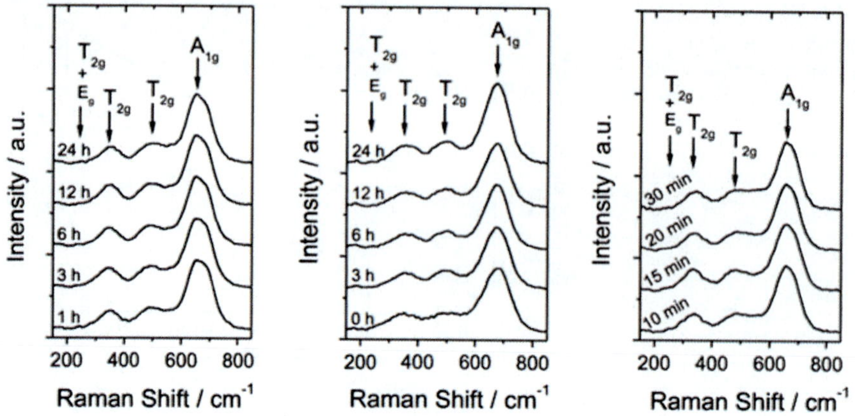

Fig. 2.4 Non-polarized Raman spectra of zinc ferrites nanoparticles prepared by hydrothermal (left), miniemulsion + hydrothermal (middle), and microwave-assisted sol–gel methods (right) after different preparation times. Reproduced from [5] with permission from The Royal Society of Chemistry

different parameters obtained via X-ray diffraction. Gomes et al. [7] are also an interesting example of the application of this technique in the determination of the inversion degrees of spinel ferrites. Zampiva et al. [8] analyzed the inversion degree of $MgFe_2O_4$ prepared via auto-combustion synthesis at different fuel-to-oxidizer ratios. They assessed the cationic arrangement by the ratio between the intensities of the [111] and [311] reflections; the smaller this value, the larger the inversion.

Raman spectroscopy has also been often used as an indicator of the cation positioning inside the spinel lattice. The spinel space group (Fd-3 m) has 5 Raman-active vibrational modes: A_{1g}, E_g, and 3 T_{2g} modes. Nevertheless, these considerations are only valid for a normal spinel structure. The cation exchange between tetrahedral and octahedral voids results in an overall decrease in symmetry. Graves, Johnston, and Campaniello [9] have published a thorough study regarding the Raman spectra of such spinels when submitted to symmetry reduction operations. This lower symmetry can give rise to different vibrational modes since the structure is not strictly a spinel anymore. The new, low-symmetry framework may, for instance, result in a split in the primary A_g signal, and the relative intensity between this novel band and the original phonon is a good indicator of the inversion degree of the system. Granone et al. [10] studied the effect of the inversion degree on the Raman spectra of zinc ferrites, while also showing a comparison among the data obtained via different characterization techniques. Their results are reproduced in Fig. 2.5.

Nevertheless, Mössbauer spectroscopy is by far the best-suited technique for the determination of inversion degrees of ferrites. This technique allows for a very precise definition of the occupation of each of the different sites, usually based on the position of the iron atoms. Mossbauer spectroscopy is considered a substitute for nuclear magnetic resonance in the determination of the structure of magnetic materials, which

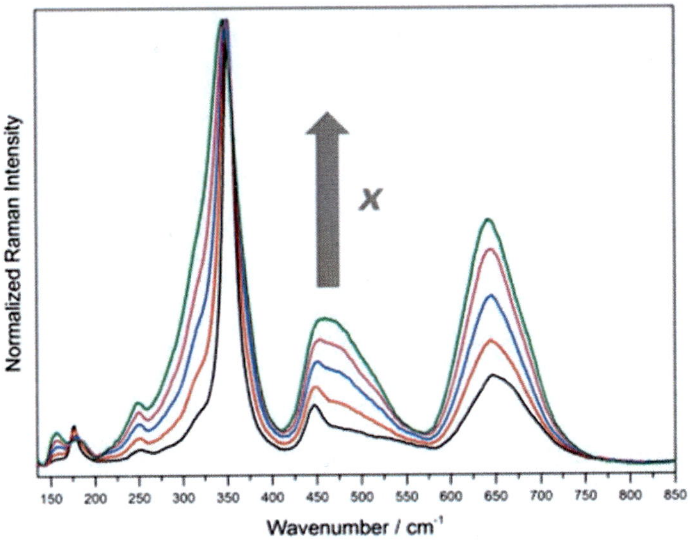

Fig. 2.5 Normalized Raman spectra of $ZnFe_2O_4$ samples with increasing degree of inversion x (ranging from 0.074 to 0.203) obtained using a 785 nm laser as the excitation source (Adapted from Granone et al. [10], published by the PCCP Owner Societies under CC BY-NC 3.0)

cannot be analyzed by the latter due to the intense magnetic fields required by this technique. Mössbauer spectroscopy yields data that can be mathematically decomposed into contributions from each cationic site. The area of each contribution can be directly related to the occupation of the sites. The inversion degree can therefore be obtained in a very straightforward manner.

Several works related to the definition of the cationic arrangement of ferrites rely heavily on their Mössbauer spectra. Venturini et al. [11] have studied the effects of different additives on the sol–gel synthesis of $CoFe_2O_4$. The various organic compounds affected the stoichiometry of the produced materials, as well as their inversion degrees. The latter was analyzed mainly via Mössbauer spectroscopy, considering the presence of 4 different sites: octahedral, tetrahedral, iron in hematite, and superparamagnetic iron. The results are reprinted in Fig. 2.6.

More complex iron distributions can also be analyzed via the Mössbauer technique. In their work from 2019, Avancini et al. [12] have prepared magnetite (Fe_3O_4) via a glass–ceramic method. The final product consisted of ferrite nanoparticles dispersed in an iron-rich glassy matrix. Despite the complexity of the system, the authors have managed to analyze the different iron populations via a careful analysis of the Mössbauer spectra of the samples. Four iron species were postulated: two sites in the spinel lattice and two populations contained in the matrix. As an example of more complex situations, the Mössbauer spectra of these glass ceramics containing different initial amounts of iron are shown in Fig. 2.7.

Besides the magnetization, the inversion degree plays a pivotal role in determining another property of ferrites: their critical or Curie temperature (T_C). This value is

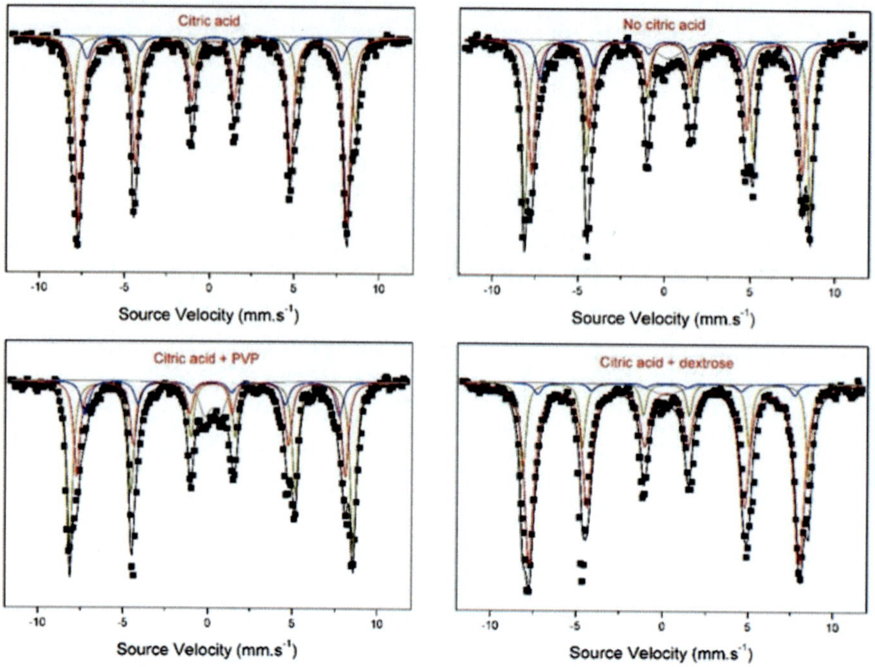

Fig. 2.6 ^{57}Fe-Mössbauer spectra of the $CoFe_2O_4$ particles prepared via the sol–gel method in the presence of different additives. Squares indicate the experimental data, while the solid black line denotes the result of the fitting procedure. Colored lines indicate the contribution from each of the four iron sites (Reprinted from [11] with permission from Elsevier)

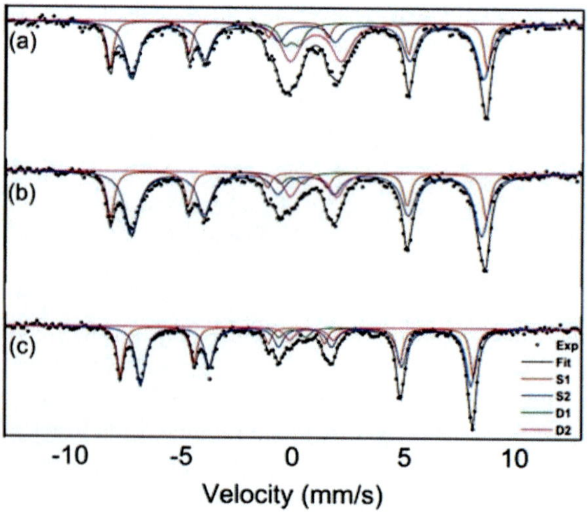

Fig. 2.7 Mössbauer spectra of iron-rich glass ceramics with (a) 20, (b) 30, and (c) 45 wt% of iron scale heat-treated at 700 °C. The colored lines represent the contributions of the different iron sites. Reprinted from [12] with permission from Elsevier

the point in which the crystal has enough thermal energy to overcome the natural tendency toward magnetic alignment of its constituent moments. Above this temperature, the ferrite transitions into a paramagnetic state. To understand how the inversion affects T_C, one must recall why the moments were spontaneously aligned to begin with. In more simple terms, the magnetization along one direction leads the system to a lower energy state, in which the ferrite is said to be relaxed. In the case of ferrites, the exchange integrals involved in the AB interaction (octahedral-tetrahedral, mediated by an O^{2-} anion) are those that most intensely determine whether the material will be spontaneously magnetized. Furthermore, the strength of this exchange defines not only the magnetization itself but also the extent to which this mechanism can withstand thermal inputs, which work toward creating random magnetic alignment.

As previously discussed, the strongest exchange interactions in ferrite systems are usually those found in Fe^{3+}(A)-Fe^{3+}(B) interactions, that is, the exchange between trivalent iron atoms in the tetrahedral and octahedral sublattices. Therefore, a completely inverted spinel would tend to show the strongest exchange interactions, which would—theoretically—lead to strong resistance against thermal demagnetization. This resistance leads to a higher critical temperature for the analyzed ferrite. In normal frameworks, the divalent cation usually shows poorer superexchange interactions. For instance, Co(A)-Fe(B) and Mn(A)-Fe(B) display superexchange interactions 68 and 66% as strong as that of Fe(A)-Fe(B) arrangements [13]. Consequently, lower critical temperatures are expected of systems that place atoms other than iron in the tetrahedral position, i.e., spinels with low inversion degree. An earlier study by the authors [14] related the degree of inversion to the observed critical temperature. Cobalt ferrite samples with an inverse configuration showed a Curie point 85 °C higher than that displayed by the normal configuration. This phenomenon was related to the stronger exchange interactions displayed by the samples that placed most iron atoms in the tetrahedral position (inverse spinels). The variation of the normalized magnetic moment of the samples can be seen in Fig. 2.8, where the intercept of the dashed lines with the x-axis indicates the Curie temperature of the samples.

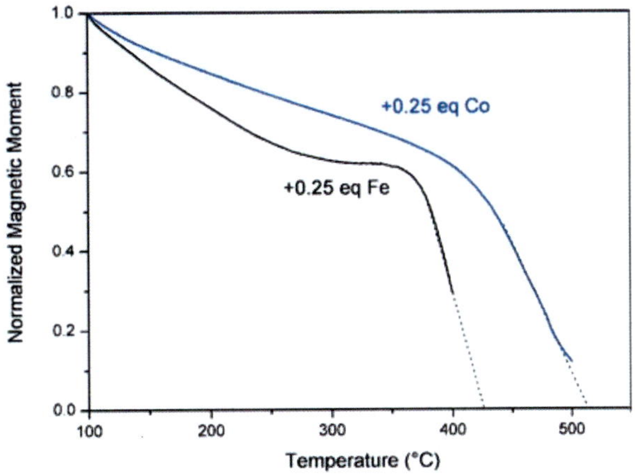

Fig. 2.8 Normalized magnetic moment of $CoFe_2O_4$ particles with normal (black line) and inverse (blue line) spinel configurations. The intercept of the dashed lines with the x-axis estimates the critical temperature of the different materials (Reprinted from [14] with permission from Elsevier)

Another good illustration of the power of the iron exchange interaction can be seen in materials where iron is selectively substituted for another trivalent cation. Compositions in which aluminum ions replaced trivalent iron are a good example thereof. Samples with an empirical formula $CoFe_{2-x}Al_xO_4$ (with x varying from 0 to 0.5) have already been described in the literature [15]. An increase in the amount of aluminum led to smaller unit cells and increased DC resistivity. More strikingly, the critical demagnetization temperature was decreased by approximately 150 °C by substituting half of the iron atoms.

Walters and Wirtz have also shown in a very ingenious study [16] the effect of the inversion degree on the Curie temperature of magnesium ferrites. Samples were quenched from different temperatures and then further treated thermally at 450 °C for varying amounts of time. This secondary treatment was intended to allow the ferrite system to relax to its most stable configuration at lower temperatures. The authors have shown that the least strained configurations (those quenched from lower temperatures) relax their cationic arrangement more rapidly than those quenched from higher temperatures. This effect can be seen in Fig. 2.9, which shows the variation of the Curie temperature of the samples submitted to ever-increasing treatment times. The results are also a very striking example of the influence of kinetic parameters on the inversion degree of ferrites.

Another exciting—and often troublesome—aspect of the spinel ferrite framework is the ease with which the structure can incorporate atomic vacancies. Indeed, the nominal formula MFe_2O_4 is very commonly found to not realistically depict the composition of the studied ferrites. These materials can comfortably accommodate substantial deviations from the expected formula, being naturally found at varying

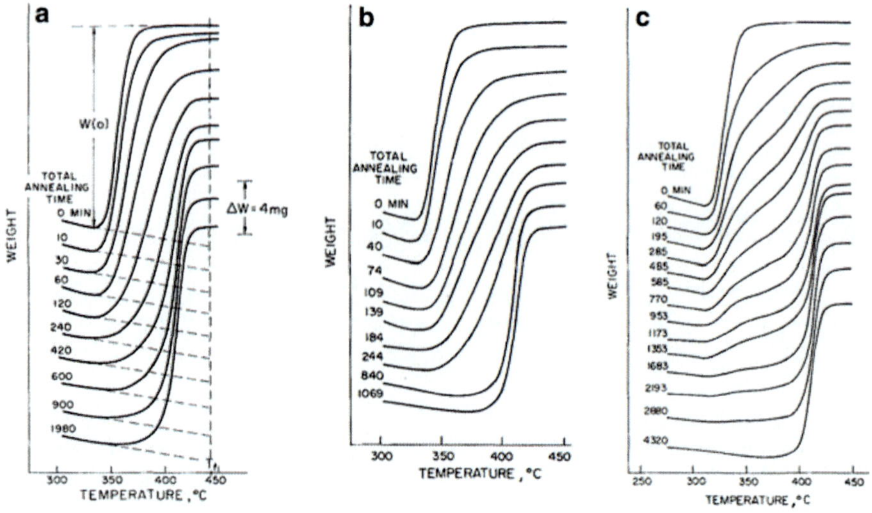

Fig. 2.9 Magnetic thermograms of $MgFe_2O_4$ quenched from (A) 1000, (B) 1150, and (C) 1295 °C after annealing at 450 °C for various times (Reprinted from [16] with permission from the American Ceramic Society)

degrees of substoichiometry. This phenomenon can be explained by the structure of iron oxides: Wüstite (FeO), hematite (-Fe_2O_3), and magnetite (Fe_3O_4) are all formed by a cubic close packing of O^{2-} anions, differing only in the positioning and charge of their cations. This character also poses a difficulty when analyzing the X-ray diffractograms of ferrites. Despite seeming to show a pure ferritic composition, many of these common impurities show reflections at the same angles as the primary oxide, thus hindering inferences over the purity of the produced ferrites. Indeed, even the superficial relaxation of ferrites (i.e., the processes via which the phase boundaries stabilize toward lower energies, such as the polar catastrophe) involves the formation of substoichiometric oxides at a subsurface level. Bliem et al. [17] have shown that Fe_3O_4 creates an ordered array of subsurface iron vacancies; they further dissert that such phenomenon occurs because these oxides prefer to redistribute cations in response to the oxidizing character of the medium in which they are inserted. Despite presenting data concerning only iron ferrite, the authors postulate that similar mechanisms might occur in other ferrites and related oxides, such as Co_3O_4.

Oxygen defects in ferrites have been widely discussed in the literature, not least because many of the properties of these ferrites are strongly affected by the presence of oxygen vacancies. Aksel'rod et al. [18] have discussed the effects of the oxygen nonstoichiometry on the phase stability of magnesium ferrite ($MgFe_2O_4$). The researchers have shown the ability of this ferrite to accommodate oxygen point defects up to 0.6 at%. Nevertheless, such capacity reaches a limit: above approximately 1300 °C, the ferrite decomposes into MgO and solid solutions of Fe_3O_4 in $MgFe_2O_4$. The decomposition of ferrites in this range of temperature is a critical topic when considering the utilization of heat treatment in the production of ferritic ceramic bodies; this topic will be further considering in the chapter concerning the processing of ferrites.

2.3 Hexaferrites

Another significant class of ferrites contains those that crystallize with hexagonal symmetry, the *hexaferrites*. Since their discovery in the 1950s, these materials have become an integral part of industry and technology in a wide variety of applications, such as magnetic recording and data storage.

Magnetoplumbite, a magnetic mineral, was first described in the 1920s. It was described as a hexagonal crystal of empirical formula $PbFe_{7.5}Mn_{3.5}Al_{0.5}O_{19}$ [19]. In the 50s, Philips Laboratories conducted several studies to reproduce a synthetic version of this mineral ($PbFe_{12}O_{19}$) and its analogs, kickstarting the nascent field of hexaferrite development. In these synthetic ferrites, lead was substituted for large divalent cations, namely strontium ($SrFe_{12}O_{19}$) and barium ($BaFe_{12}O_{19}$). Today, this class of compounds represents by far the most economically important magnetic materials.

Table 2.2 Subgroups of hexaferrites, their chemical formula, and examples of each class, with their usual nomenclature

Subgroup	Formula	Example	Nomenclature
M-type	$MeFe_{12}O_{19}$	$SrFe_{12}O_{19}$	SrM
Y-type	$Ba_2Me_2Fe_{12}O_{22}$	$Ba_2Co_2Fe_{12}O_{22}$	Co_2Y
W-type	$BaMe_2Fe_{16}O_{27}$	$BaCo_2Fe_{16}O_{27}$	Co_2W
Z-type	$Ba_3Me_2Fe_{24}O_{41}$	$Ba_3Co_2Fe_{24}O_{41}$	Co_2Z
X-type	$Ba_2Me_2Fe_{28}O_{46}$	$Ba_2Co_2Fe_{28}O_{46}$	Co_2X
U-type	$Ba_4Me_2Fe_{36}O_{60}$	$Ba_4Co_2Fe_{36}O_{60}$	Co_2U

As do their spinel counterparts, hexaferrites are composed of oxygen, iron, and at least one other metallic atom. Barium and strontium are the most common elements present in these materials. Unlike spinels, however, hexaferrites cannot be easily represented by a single space group or even a simplified formula. Instead, this large family of ferrites is subdivided into several groups with similar crystallography and composition. The most common ones are represented in Table 2.2.

At a chemical level, hexaferrites can be understood as combinations of spinel ferrites and Y and M ferrites. Z ferrites ($Ba_3Me_2Fe_{24}O_{41}$), for instance, may be represented as a sum of the corresponding Y ($Ba_2Me_2Fe_{12}O_{22}$) and M ferrite ($BaFe_{12}O_{19}$). Crystallographically, all hexaferrites can be understood as composed of spinel and hexagonal layers. Different stacking arrangements of these basic components would create the unit cells of each subgroup of hexaferrites. Pullar [20] wrote a comprehensive description of the different stacking possibilities in these systems. An example of this layered structure in the unit cell of barium ferrite can be seen in Fig. 2.10.

As would be expected, the hexaferrite families do not differ only in their crystallographic framework. This variety of stacking possibilities leads to major effects on their magnetic properties. The large magnetocrystalline anisotropy of hexaferrites is strongly dependent on the family to which they belong.

M ferrites were the first described hexaferrites and are still the most economically relevant. SrM and BaM are the two most common compounds associated with this family. Ferrites in this group crystallize in the $P6_3/mmc$ space group, as do W and Z ferrites. The remaining families (Y, X, and U) belong to the R3c group. Their composition—as well as the hexagonal framework—usually leads to large magnetocrystalline anisotropies aligned in the direction of the c-axis. Large anisotropies translate into very high-energy requirements for the movement of the magnetization direction. The macroscopic consequence of this phenomenon is their elevated coercivity, one of the most striking features of hexaferrites. As a bonus, the high Curie temperature of these materials allows their application at room temperature.

Besides uniaxial anisotropy, hexaferrites may also show magnetization along a plane, usually perpendicular to the c-axis. Many Y ferrites, for instance, show this type of magnetic behavior, as is the case of Co_2Y [23, 24]. Hexaferrites may also show a "cone" of anisotropy. In this case, the easy axis is not entirely fixed but has some angular value in which the material can be magnetized without energy expenditure. This angle must be smaller than 90°. At room temperature, $BaCo_2Fe_{16}O_{27}$ or Co_2W displays a cone of magnetization at an angle of approximately 70° with the c-axis

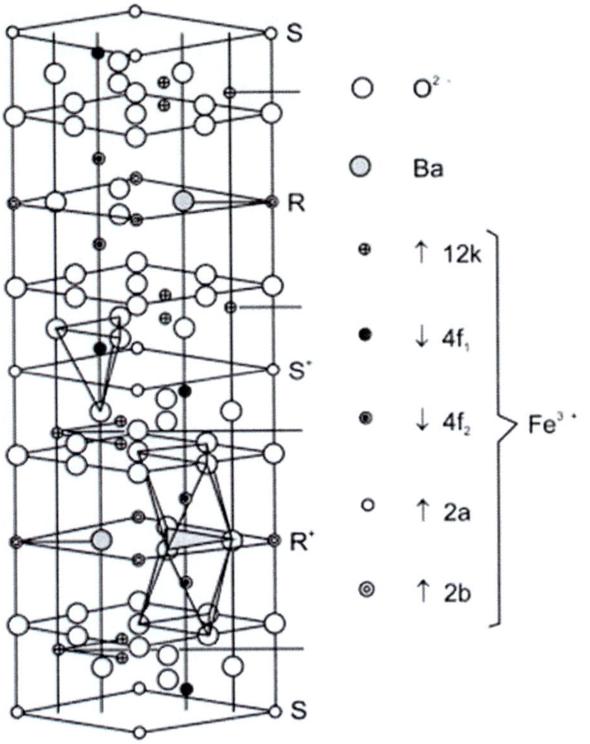

Fig. 2.10 Unit cell of $BaFe_{12}O_{19}$ (BaM). The S units represent the two oxygen layers of a spinel substructure, while R is a three-layer block containing the barium atom. The arrows show the spin orientation of each iron site. Reprinted from [21, 22] with permission from Elsevier

[25]. This compound maintains this magnetic structure up to 180 °C. Above this threshold, the angle decreases smoothly; at 280 °C, the magnetization is uniaxial up to the Curie temperature [26].

The complex crystalline structure of hexaferrites lends itself easily to substitution with different metals. Materials with five or more cations are rather common in the literature [27–29]. The sheer variety of possible ferrites means that testing each composition for improvements in specific properties is a daunting task. In this regard, ab initio simulations are proving to be a reliable tool. Dixit et al. [30] screened 21 possible substituents at the iron site of barium hexaferrite. The authors performed DFT simulations of $SrFe_{11.5}M_{0.5}O_{19}$ structures and analyzed the properties of the final compounds. Out of the 21 assessed elements, only 12 showed negative substitution energies. The results showed that Bi, Sb, Sn, and Sc should induce increases in saturation magnetization. On the other hand, P, Co, Al, Ga, and Ti have shown to theoretically enhance coercivity. The results are valuable in the search for novel materials for advanced applications.

Some hexaferrites have been the focus of considerable research due to their coupling between magnetic and electric properties. Materials presenting more than one ferroic property (e.g., ferroelectricity, ferromagnetism) are said to be multiferroic. The coupling between electric and magnetic character opens up the possibility of creating novel devices in which magnetic fields are used to affect the electrical properties of the material or, conversely, electric fields induce changes in the magnetization of the device. Several Z and Y ferrites, such as $Ba_2Mg_2Fe_{12}O_{22}$ and $Sr_3Co_2Fe_{24}O_{41}$, have already shown multiferroic character [31, 32]. $Sr_3Co_2Fe_{24}O_{41}$, which shows this type of coupling at room temperature, has already been successfully applied as a material in prototypical four-state memory devices [33].

2.4 Perovskite Ferrites

The third great family of ferrites is the perovskite ferrites, also known as orthoferrites. Although less varied than the two previous classes, some perovskite ferrites have been intensely researched due to their unique properties. Two of these compounds will be discussed in this section, bismuth ferrite and lanthanum strontium cobalt ferrite.

Perovskite ferrites are materials with a chemical formula $MeFeO_3$ (in which Me is a trivalent cation) with crystal structures pertaining to the perovskite series. Figure 2.11 shows the possible crystalline arrangements of bismuth ferrite ($BiFeO_3$), which is perhaps the most studied example of this class of ferrites.

At room temperature, bismuth ferrite presents a rhombohedral arrangement (R3c). This material exhibits natural electric polarization along the <111> pseudocubic direction with a ferroelectric Curie temperature of 1043 K. The magnitude of this natural polarization may reach over 100 $\mu C.cm^{-2}$ [35]. However, the presence of oxygen defects may drastically decrease the intensity of this effect [36]. Excessive current leakage in site defects is observed in $BiFeO_3$, even at low temperatures. Therefore, stoichiometry control is fundamental in the production of these multiferroics.

Bismuth ferrite also presents antiferromagnetism, with a Neel temperature of approximately 650 K. It is a G-type antiferromagnet, in which each iron spin is aligned antiparallel to its neighboring Fe^{3+} spins. In fact, weak spin canting is observed due to local magnetic coupling to the polarization of the structure. Hence, the canted magnetism and ferroelectricity of $BiFeO_3$ at room temperature make this material very interesting for multiferroics applications. One of the areas in which this material is intensively studied is spintronics, which is a novel field of electronics that utilizes electronic spin as a further degree of freedom. This type of informatics should have profound effects on the processing speed and capacity of novel solid-state devices. Multiferroic materials are applied in spintronics, for instance, in the production of low-power electrically controlled spintronic devices, such as spin valves [37]. In this regard, the utilization of $BiFeO_3$ proves useful when trying to control the electrical properties of a device via the application of a magnetic bias, or vice versa [38]. An example is shown in, which shows the effect on the magnetoresistance of a spin valve after the application of different voltage pulses to a $BiFeO_3$/$BiFeO_3$-Mn bilayer (Fig. 2.12).

Another material worth mentioning in this chapter is lanthanum strontium cobalt ferrites (LSCF). These compounds of general formula $La_{1-x}Sr_xCo_yFe_{1-y}O_{3-\delta}$ undergo

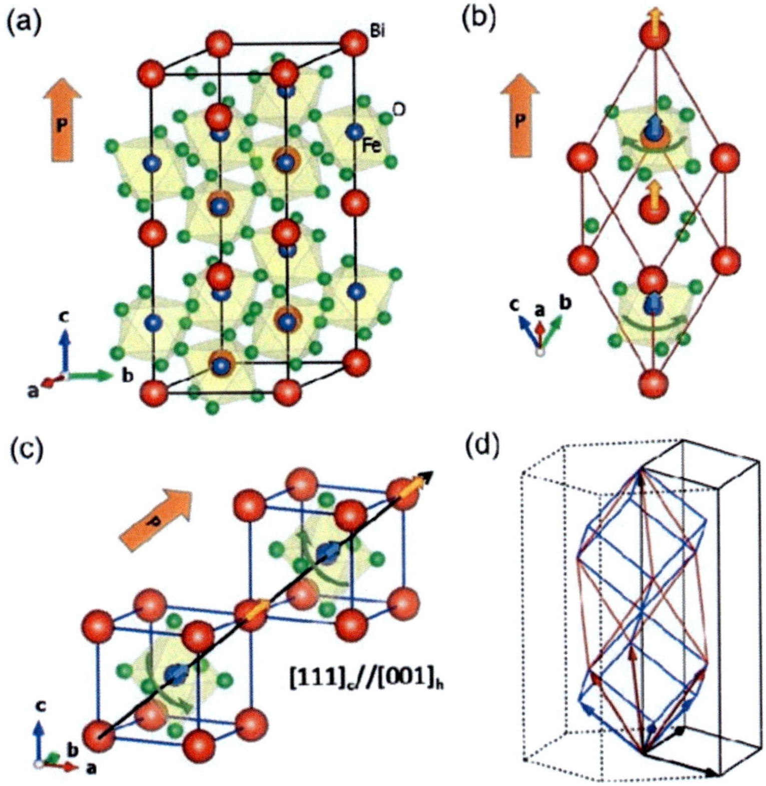

Fig. 2.11 (a) Hexagonal, (b) rhombohedral, and (c) pseudocubic unit cells of $BiFeO_3$. (d) Unit vectors and unit cell in the hexagonal (black), rhombohedral (red), and pseudocubic (blue) notations (Reprinted from [34] with permission from IOP Publishing)

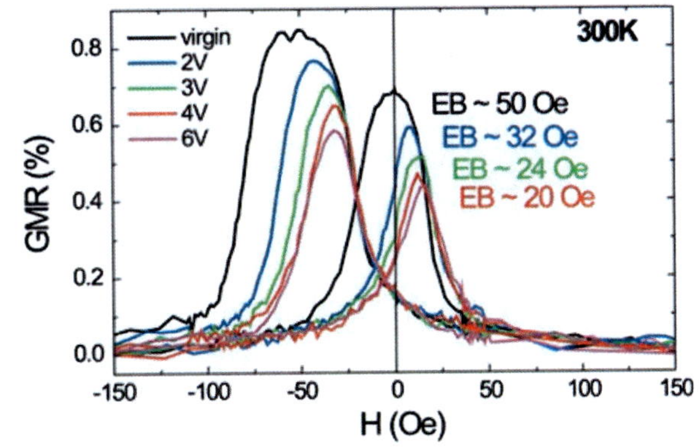

Fig. 2.12 Giant magnetoresistance of Au/Co/Cu/CoFeB spin valve after different voltage pulses applied to a $BiFeO_3$-based film (Reprinted from [37] with permission from the American Chemical Society)

several structural transitions associated with the perovskite series. LSCF is perhaps one of the most studied materials for cathodes of solid oxide fuel cells (SOFCs), owing to their mixed ionic and electronic conduction [39]. This material usually presents electronic and ionic conductivities at 800 °C on the order of 10^2 and 10^{-1} $S.cm^{-1}$, respectively [40]. Design advances and improved formulations may yield even higher values. Banerjee et al. [40] have shown that LSCF/lanthanum iron cobaltite composite cathodes may reach 686 $S.cm^{-1}$ of electronic conductivity. The utilization of electrospun LCSF fibers also shows great promise for application in intermediate-temperature fuel cells [41].

2.5 Garnet Ferrites

The last family of ferrites to be discussed in this chapter are the garnet ferrites. These materials, of chemical formula $Me_3Fe_5O_{12}$ or $3Me_2O_3.5Fe_2O_3$, belong to the Ia-3d space group, meaning they crystallize in the dodecahedral structures associated with the mineral garnet. The substituent trivalent metal is usually a rare-earth element. Yttrium, the structural component of the most famous garnet ferrite, is not a rare-earth element per se. Still, it has properties very similar to those of lanthanides, so it is accordingly grouped with the latter.

There are three types of cationic sites in the garnet unit cell: 16 octahedral, 24 tetrahedral, and 24 dodecahedral, as depicted in Fig. 2.13. Only the dodecahedral

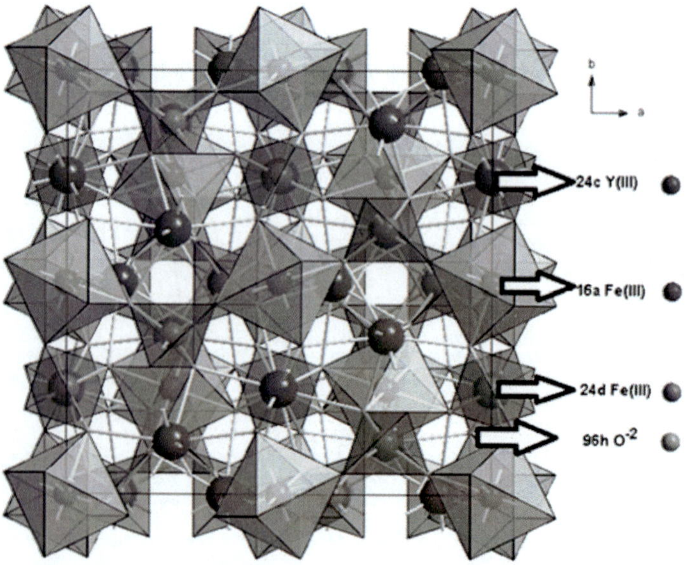

Fig. 2.13 Representation of the cubic unit cell of $Y_3Fe_5O_{12}$. Reprinted from [42] with permission from Springer

sites are large enough to contain the rare-earth. Therefore, garnet ferrites present much less structural variation than their spinel relatives. In terms of exchange interactions between the sublattices, all octahedral and dodecahedral cations are aligned antiparallel to the Fe^{3+} ions in the tetrahedral site. If the substituent is a lanthanide, extra spin contribution from this element will also be responsible for the magnetism of the material. In $Y_3Fe_5O_{12}$, also commonly known as YIG (yttrium iron garnet), the resulting moment after mutual cancelation comes from tetrahedral Fe^{3+}. Considering a single unit cell, the contribution to magnetization comes from 8 out of the 24 iron atoms in tetrahedral sites or a fifth of the total iron content. If a lanthanide is present in the structure, some orbital contribution to the magnetic moment is also expected.

YIG is understandably the most utilized garnet ferrite. It is a ferrimagnet with a Curie temperature of approximately 560 K. The magnetic character above room temperature, coupled to an extremely narrow ferromagnetic resonance line and low spin-wave damping, makes YIG an ideal material for studies on the physics of magnets [43]. Besides these more academic purposes, YIG is also widely used in microwave devices, such as bandpasses or bandstops. YIG spheres are one of these common applications. In these filters, which have typical Q factors on the order of 10^2, the resonant frequency of a garnet sphere is tuned via the action of an electromagnet. The range of magnetic fields applied on the sphere translates into different frequencies being filtered off or allowed through the system.

YIG is also essential in the field of magnonics. Magnons are spin-wave excitations in magnetically ordered media that may be engineered for advanced data storage. YIG exhibits magnetic damping about two orders of magnitude lower than that of ferromagnetic metals; this is perhaps the main reason behind the choice of this garnet as the foremost material in magnonics [44]. Furthermore, yttrium iron garnet films also show photon-magnon coupling [45], which indicates that these magnonic devices may use light as a means of transferring information to the storage system. Bhoi et al. [46] recently demonstrated robust photon-magnon coupling in YIG films and microwave photons in an inverse pattern of split-ring resonator. The authors report a normalized coupling strength of 0.194 Hz at 3.7 GHz.

References

1. Faller JG, Birchenall CE (1970) The temperature dependence of ordering in magnesium ferrite. J Appl Crystallogr 3:496–503. https://doi.org/10.1107/S0021889870006751
2. O'neill HSC (1992) Temperature dependence of the cation distribution in zinc ferrite (ZnFe2O4) from powder XRD structural refinements. Eur J Mineral 4:571–580. https://doi.org/10.1127/ejm/4/3/0571.
3. Šepelák V, Becker KD (2004) Comparison of the cation inversion parameter of the nanoscale milled spinel ferrites with that of the quenched bulk materials. Mater Sci Eng A 375–377:861–864. https://doi.org/10.1016/j.msea.2003.10.178
4. Šepelák V, Schultze D, Krumeich F, Steinike U, Becker K (2001) Mechanically induced cation redistribution in magnesium ferrite and its thermal stability. Solid State Ionics 141–142:677–682. https://doi.org/10.1016/S0167-2738(01)00777-9

5. Dolcet P, Kirchberg K, Antonello A, Suchomski C, Marschall R, Diodati S, Muñoz-Espí R, Landfester K, Gross S (2019) Exploring wet chemistry approaches to ZnFe2O4 spinel ferrite nanoparticles with different inversion degrees: a comparative study. Inorg Chem Front 6:1527–1534. https://doi.org/10.1039/C9QI00241C
6. Podwórny J (2015) Temperature dependence of the inversion degree in three-cation spinel solid solutions: experimental evaluation by XRD. Powder Diffr 30:S104–S110. https://doi.org/10.1017/S0885715615000093
7. Gomes JA, Sousa MH, Tourinho FA, Mestnik-Filho J, Itri R, Depeyrot J (2005) Rietveld structure refinement of the cation distribution in ferrite fine particles studied by X-ray powder diffraction. J Magn Magn Mater 289:184–187. https://doi.org/10.1016/j.jmmm.2004.11.053
8. Zampiva RYS, Kaufmann CG, Alves AK, Bergmann CP (2018) Influence of the fuel composition and the fuel/oxidizer ratio on the combustion solution synthesis of MgFe2O4 catalyst nanoparticles. FME Trans 46:157–164. https://doi.org/10.5937/fmet1802157Z
9. Graves PR, Johnston C, Campaniello JJ (1988) Raman scattering in spinel structure ferrites. Mater Res Bull 23:1651–1660. https://doi.org/10.1016/0025-5408(88)90255-3
10. Granone LI, Ulpe AC, Robben L, Klimke S, Jahns M, Renz F, Gesing TM, Bredow T, Dillert R, Bahnemann DW (2018) Effect of the degree of inversion on optical properties of spinel ZnFe 2 O 4. Phys Chem Chem Phys 20:28267–28278. https://doi.org/10.1039/C8CP05061A
11. Venturini J, Zampiva RYS, Arcaro S, Bergmann CP (2018) Sol-gel synthesis of substoichiometric cobalt ferrite (CoFe2O4) spinels: influence of additives on their stoichiometry and magnetic properties. Ceram Int 44:12381–12388. https://doi.org/10.1016/J.CERAMINT.2018.04.026
12. Avancini TG, Souza MT, de Oliveira APN, Arcaro S, Alves AK (2019) Magnetic properties of magnetite-based nano-glass-ceramics obtained from a Fe-rich scale and borosilicate glass wastes. Ceram Int 45:4360–4367. https://doi.org/10.1016/J.CERAMINT.2018.11.111
13. Sawatzky GA, Van Der Woude F, Morrish AH (1969) Mössbauer study of several ferrimagnetic spinels. Phys Rev 187:747–757. https://doi.org/10.1103/PhysRev.187.747
14. Venturini J, Tonelli AM, Wermuth TB, Zampiva RYS, Arcaro S, Da Cas Viegas A, Bergmann CP (2019) Excess of cations in the sol-gel synthesis of cobalt ferrite (CoFe2O4): A pathway to switching the inversion degree of spinels. J Magn Magn Mater 482:1–8. https://doi.org/10.1016/j.jmmm.2019.03.057
15. Gul IH, Ahmed W, Maqsood AÃ (2008) Electrical and magnetic characterization of nanocrystalline Ni – Zn ferrite synthesis by co-precipitation route. J Magn Magn Mater 320:270–275. https://doi.org/10.1016/j.jmmm.2007.05.032
16. Walters DS, Wirtz GP (1971) Kinetics of cation ordering in magnesium ferrite. J Am Ceram Soc 54:563–566. https://doi.org/10.1111/j.1151-2916.1971.tb12208.x
17. Bliem R, McDermott E, Ferstl P, Setvin M, Gamba O, Pavelec J, Schneider MA, Schmid M, Diebold U, Blaha P, Hammer L, Parkinson GS (2014) Subsurface cation vacancy stabilization of the magnetite (001) surface. Science (80-.):346. https://doi.org/10.1126/science.1260556.
18. Aksel'rod EI, Alapin BG, Vishnevsky II, Sukharevsky BY (1971) Structural and phase relations in nonstoichiometric ferrites with oxygen deficiencies. J Phys Chem Solids 32:1627–1639. https://doi.org/10.1016/S0022-3697(71)80056-2
19. Collongues R, Gourier D, Kahn-Harari A, Lejus AM, Thery J, Vivien D (1990) Magnetoplumbite-related oxides. Annu Rev Mater Sci 20:51–82. https://doi.org/10.1146/annurev.ms.20.080190.000411
20. Pullar RC (2012) Hexagonal ferrites: a review of the synthesis, properties and applications of hexaferrite ceramics. Prog Mater Sci 57:1191–1334. https://doi.org/10.1016/j.pmatsci.2012.04.001
21. Mariño-Castellanos PA, Moreno-Borges AC, Orozco-Melgar G, García JA, Govea-Alcaide E (2011) Structural and magnetic study of the Ti4+-doped barium hexaferrite ceramic samples: Theoretical and experimental results. Phys B Condens Matter 406:3130–3136. https://doi.org/10.1016/j.physb.2011.03.084
22. Wartewig P, Krause MK, Esquinazi P, Rösler S, Sonntag R (1999) Magnetic properties of Zn- and Ti-substituted barium hexaferrite. J Magn Magn Mater 192:83–99. https://doi.org/10.1016/S0304-8853(98)00382-5

23. Bai Y, Zhou J, Gui Z, Li L (2002) Magnetic properties of Cu, Zn-modified Co2Y hexaferrites. J Magn Magn Mater 246:140–144. https://doi.org/10.1016/S0304-8853(02)00040-9
24. Tchouank Tekou Carol T, Mohammed J, Basandrai D, Godara SK, Bhadu GR, Mishra S, Aggarwal N, Narang SB, Srivastava AK (2020) X-band shielding of electromagnetic interference (EMI) by Co2Y barium hexaferrite, bismuth copper titanate (BCTO), and polyaniline (PANI) composite. J Magn Magn Mater 501:166433. https://doi.org/10.1016/j.jmmm.2020.166433
25. Samaras D, Collomb A, Hadjivasiliou S, Achilleos C, Tsoukalas J, Pannetier J, Rodriguez J (1989) The rotation of the magnetization in the BaCo2Fe16O27 W-type hexagonal ferrite. J Magn Magn Mater 79:193–201. https://doi.org/10.1016/0304-8853(89)90098-X
26. Mahmood SH, Al Sheyab Q, Bsoul I, Mohsen O, Awadallah A (2018) Structural and magnetic properties of Ga-substituted Co2−W hexaferrites. Curr Appl Phys 18:590–598. https://doi.org/10.1016/j.cap.2018.02.013
27. Alizad Farzin Y, Mirzaee O, Ghasemi A (2014) Influence of Mg and Ni substitution on structural, microstructural and magnetic properties of Sr2Co2−xMgx/2Nix/2Fe12O22 (Co2Y) hexaferrite. J Magn Magn Mater 371:14–19. https://doi.org/10.1016/j.jmmm.2014.07.007
28. Wu M, Gao X, Liu Z (2015) Field-dependent magnetoelectric effects in polycrystalline Co2Y-Type Ba0.5Sr1.5Co2(Fe1−xAlx)12O22 hexaferrites. J Am Ceram Soc 98:2498–2502. https://doi.org/10.1111/jace.13637
29. Vinaykumar R, Jyoti J (2018) Bera, characterization of La-Zn substituted Co2Y hexagonal ferrite. J Electron Mater 47:5959–5964. https://doi.org/10.1007/s11664-018-6486-0
30. Dixit V, Kim S-G, Park J, Hong Y-K (2017) Effect of ionic substitutions on the magnetic properties of strontium hexaferrite: a first principles study. AIP Adv 7:115209. https://doi.org/10.1063/1.4995309
31. Komandin GA, Prokhorov AS, Torgashev VI, Zhukova ES, Gorshunov BP, Bush AA (2011) Polarization modes in the Ba2Mg2Fe12O22 multiferroic. Phys Solid State 53:736. https://doi.org/10.1134/S1063783411040202
32. Pullar RC (2014) Multiferroic and magnetoelectric hexagonal ferrites BT—Mesoscopic phenomena in multifunctional materials: synthesis, characterization, modeling and applications. In: Saxena A, Planes A (eds), Springer Berlin Heidelberg, Berlin, Heidelberg, pp 159–200. https://doi.org/10.1007/978-3-642-55375-2_7.
33. Wu J, Shi Z, Xu J, Li N, Zheng Z, Geng H, Xie Z, Zheng L (2012) Synthesis and room temperature four-state memory prototype of Sr3Co2Fe24O41 multiferroics. ApplPhys Lett 101:122903. https://doi.org/10.1063/1.4753973
34. Park J-G, Le MD, Jeong J, Lee S (2014) Structure and spin dynamics of multiferroic BiFeO3. J PhysCondens Matter 26:433202. https://doi.org/10.1088/0953-8984/26/43/433202
35. Lebeugle D, Colson D, Forget A, Viret M (2007) Very large spontaneous electric polarization in BiFeO3 single crystals at room temperature and its evolution under cycling fields. Appl Phys Lett 91:22907. https://doi.org/10.1063/1.2753390
36. Dho J, Qi X, Kim H, MacManus-Driscoll JL, Blamire MG (2006) Large electric polarization and exchange bias in multiferroic BiFeO3. Adv Mater 18:1445–1448. https://doi.org/10.1002/adma.200502622
37. Allibe J, Fusil S, Bouzehouane K, Daumont C, Sando D, Jacquet E, Deranlot C, Bibes M, Barthélémy A (2012) Room temperature electrical manipulation of giant magnetoresistance in spin valves exchange-biased with BiFeO3. Nano Lett 12:1141–1145. https://doi.org/10.1021/nl202537y
38. Zhang X, Wang YH, Zhang DL, Zhang GQ, Yang HL, Miao J, Xu XG, Jiang Y (2011) Electric-field-induced change of the magnetoresistance in the multiferroic spin-valve based on ${\hbox{BiFeO}}_{3}$ Film. IEEE Trans Magn 47:3139–3142. https://doi.org/10.1109/TMAG.2011.2148102
39. Jiang SP (2002) A comparison of O2 reduction reactions on porous (La,Sr)MnO3 and (La,Sr)(Co,Fe)O3 electrodes. Solid State Ionics 146:1–22. https://doi.org/10.1016/S0167-2738(01)00997-3

40. Jiang SP (2019) Development of lanthanum strontium cobalt ferrite perovskite electrodes of solid oxide fuel cells – A review. Int J Hydrogen Energy 44:7448–7493. https://doi.org/10.1016/j.ijhydene.2019.01.212
41. Lubini M, Chinarro E, Moreno B, Jurado JR, de Sousa VC, Alves AK, Ribeiro JLD, Bergmann CP (2017) Electrochemical characteristics of La0.6Sr0.4Co1−yFeyO3 (y=0.2–1.0) fiber cathodes. Ceram Int 43:8715–8720. https://doi.org/10.1016/j.ceramint.2017.04.002
42. Barrón JF, Montiel H, Gómez-Vidales V, Conde-Gallardo A, Alvarez G (2017) YIG films through synthesis by means of the polymeric precursor method: correlation between the structural and vibrational properties with magnetic behavior. J Supercond Nov Magn 30:2515–2522. https://doi.org/10.1007/s10948-017-4020-x
43. Cherepanov V, Kolokolov I, L'vov V (1993) The saga of YIG: spectra, thermodynamics, interaction and relaxation of magnons in a complex magnet. Phys Rep 229:81–144. https://doi.org/10.1016/0370-1573(93)90107-O
44. Qin H, Both G-J, Hämäläinen SJ, Yao L, van Dijken S (2018) Low-loss YIG-based magnonic crystals with large tunable bandgaps. Nat Commun 9:5445. https://doi.org/10.1038/s41467-018-07893-5
45. Bhoi B, Cliff T, Maksymov IS, Kostylev M, Aiyar R, Venkataramani N, Prasad S, Stamps RL (2014) Study of photon–magnon coupling in a YIG-film split-ring resonant system. J Appl Phys 116:243906. https://doi.org/10.1063/1.4904857
46. Bhoi B, Kim B, Kim J, Cho Y-J, Kim S-K (2017) Robust magnon-photon coupling in a planar-geometry hybrid of inverted split-ring resonator and YIG film. Sci Rep 7:11930. https://doi.org/10.1038/s41598-017-12215-8

Chapter 3
Magnetic, Electrical, and Optical Properties of Ferrites

This chapter deals with the magnetic, electrical, and optical properties of spinel ferrites and their relation with the chemical composition and structural arrangements of these materials. The main parameters of a magnetic hysteresis curve are discussed, and common values for the different families of ferrites are presented. The effect of smaller crystallite sizes is explained, and the mechanism for power dispersion under an alternating magnetic field is discussed. Electron hopping mechanisms are analyzed with regard to their temperature dependence. Metal-to-insulator transitions present in spinel ferrites are demonstrated, with a particular focus on the Verwey transition. The optical properties of these materials are discussed along with their diffuse reflectance spectra. The band gaps of selected ferrites and examples of their band structure and quantum efficiency curves are demonstrated.

3.1 Introduction

The electronic configuration of a material defines the majority of its physical properties. Particularly in the case of spinel ferrites, the existence of unpaired electrons and the consequent net magnetic moment generate peculiar behaviors. The crystalline arrangement of spinel ferrites—discussed in the previous chapter—is composed of a face-centered cubic structure of oxygen ions with the tetrahedral and octahedral voids partially filled by Fe^{3+} and divalent cations. The unpaired spins in each of the sublattices are aligned antiparallel with regard to the net moment of the opposite sublattice. Furthermore, the distribution of divalent and trivalent cations is not fixed, which results in a wide possibility of electronic arrangements for crystals with the same chemical composition. The effects of the flexible structural and electronic characters of spinel ferrites are felt most intensely in the properties directly related to their band structure, namely the magnetic, electrical, and optical properties.

S. Arcaro and J. Venturini, *Modern Ferrites in Engineering*, Topics in Mining, Metallurgy and Materials Engineering, https://doi.org/10.1007/978-3-030-78988-6_3

3.2 Magnetism

The most remarkable property of spinel ferrites is indeed their magnetism. As discussed in the previous section (Chapter 2), this property is a direct consequence of the spin orientation of the different metallic ions in the spinel structure. Atoms in the tetrahedral sublattice can align their spins either parallel or antiparallel to the direction of the octahedral ions via exchange interactions mediated by oxygen anions. Generally, the sublattices are aligned in an antiparallel manner, giving rise to a ferrimagnetic alignment when the magnetic moments of the cations do not cancel each other out.

One way of analyzing magnetic properties is by obtaining the hysteresis curves of the studied material. An example of a typical hysteresis behavior is shown in Fig. 3.1. These assays allow for the extraction of several magnetic parameters. Coercivity (Hc), remanence (Mr), and saturation magnetization (Ms) are the most important among these.

Remanence or remanent magnetization refers to the magnetization of the sample in the absence of an applied field. This value is obtained at the point where the hysteresis curve crosses the *y* axis. This parameter denotes the amount of spontaneous magnetization present in a material at the temperature in which the experiment was conducted.

Saturation magnetization is the maximal magnetization that can be achieved by a material when it is submitted to strong magnetic fields. Since this parameter denotes a maximum in the magnetism shown by the material, it is usually a good indication of the internal configuration of the cations inside the crystalline framework. Saturation is only said to be achieved when the magnetization does not increase with further increments in the strength of the applied magnetic field. In the case of incomplete

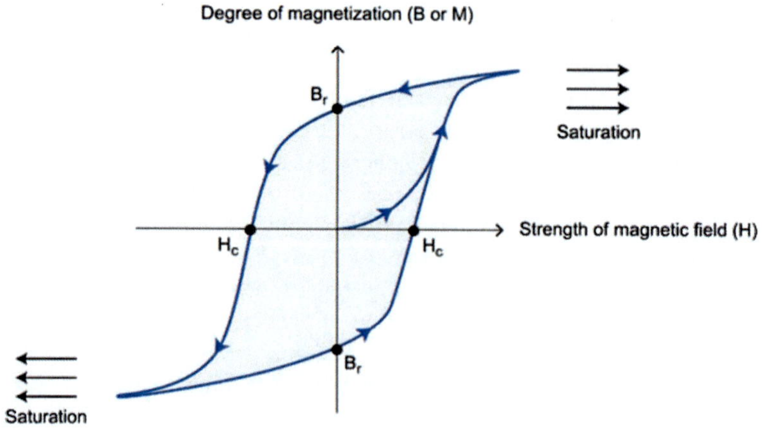

Fig. 3.1 Example of a typical magnetic hysteresis curve

saturation, maximal magnetization (M*) might be a more appropriate terminology for the parameter.

The coercivity of a sample indicates the field necessary for the inversion of the magnetization direction. In other terms, this parameter is the field in which the hysteresis curve crosses the x axis of Fig. 3.1. Coercivity is usually related to the difficulty in moving the magnetic domain walls inside a crystal. Therefore, any process or property that might hinder or facilitate the movement of these domain walls will have a direct effect on the coercivity of the material. A plethora of factors might affect this parameter. Structural factors, such as crystallite size and the presence of defects, are well regarded as capable of having a strong influence on the energy necessary for the motion of domain walls. Magnetic anisotropies (shape, stress, magnetocrystalline, among others) also define the final intensity of the coercive field. The value of coercivity of each ferrite leads to their characterization as a soft or hard ferrite.

Soft ferrites present low coercivities; their magnetization direction is easily flipped. Most of the spinel ferrites fall under this category. $NiFe_2O_4$, for instance, usually shows coercive fields on the order of 100 Oe [1, 2]. Higher values can also be achieved for this type of material via variation in annealing or synthesis technique. Azizi and Sadrnezhadd [3] have reported nickel ferrites with coercivities of 217 Oe via the annealing at 600 °C of nickel ferrites synthesized via a mechanical pathway. Other thermal treatments lead to coercivities of approximately 50 Oe. The authors assign the elevated coercive fields to the appearance of a pseudo-single domain in the annealed particles. Nevertheless, even these higher values are still low enough for these samples to be characterized as soft ferrites.

On the other hand, ferrites with high coercivities are denominated hard ferrites. Although there is no magic number separating ferrites into soft and hard ones, strontium hexaferrite ($SrFe_{12}O_9$) and barium hexaferrite ($BaFe_{12}O_9$) undoubtedly fall into the latter category, as do many of the hexaferrites. These materials usually display high coercivities and high remanence. Their high resistance to becoming demagnetized means they are well-suited for utilization as permanent magnets. Indeed, a large portion of the magnetic ferrites market is dominated by barium hexaferrite primarily due to this discussed character. Values on the order of a few kilooersted are typical for these materials, making them promising candidates for high-density storage media. Dho and collaborators [4] report coercive fields of approximately 4 kOe for $BaFe_{12}O_9$ synthesized via a conventional solid-state method and annealed at 1000 °C. The ultrafine barium ferrites reported by Liu et al. [5] displayed 5.6 kOe of coercivity. The materials were prepared via a water-in-oil emulsion method to obtain the nanometric, hard materials. Substitution with aluminum has also been shown to lead to even harder ferrites. $BaFe_{10}A_{12}O_{19}$ prepared via mechanical alloying and further treated at 1000 °C has shown coercivity values on the order of 10 kOe [6]. Wang et al. [7] have attained ferrites displaying coercivities close to 17 kOe via a careful choice of the amount of aluminum in the nominal formula of these barium ferrites.

Some spinel ferrites have coercivities high enough for them to be considered semi-hard or soft-hard. Chief among these is cobalt ferrite. The high magnetocrystalline anisotropy of Co^{2+} usually translates into elevated coercivities. Cobalt, under certain conditions, requires relatively high energies to invert its direction of magnetization. Hence, the overall coercivity of cobalt-containing spinel ferrites is unusually elevated. Values on the order of 2000 Oe are not uncommon in literature [8]. Cedeño-Mattei and collaborators [9] have reported cobalt ferrite nanoparticles presenting a coercivity of 4412 Oe. They attributed these elevated values to surface anisotropy, cation distribution, and size effects. Partial substitution of the divalent ion with cobalt is also one of the strategies employed when higher coercivities are required. Kambale et al. [10] have managed to increase the coercive field of nickel ferrite from 170 to 300 Oe by adding cobalt to their spinels.

Regarding the size dependence, ferrites display a tendency similar to that observed for most magnetic materials, as shown in Fig. 3.2. Starting from large crystals (bulk) decreases in particle size lead to a slight increase in coercivity. This increase reaches a maximum when the particle becomes too small to contain any domain walls. Said systems are called single domain. Any decrease in particle size after this point leads to a decline in coercivity. At its limit, this trend leads to the appearance of magnetic particles with null coercivity. These particles have thus become superparamagnetic (SPM). In this region, thermal contributions dominate the system and cause spontaneous variation in the direction of the spin alignment. Nevertheless, the particles may still demonstrate the same remanence and saturation magnetization as their bulk counterparts. Superparamagnetic behavior in ferrites usually appears at sizes of a few tens of nanometers.

In the superparamagnetic state, thermal energy alone is enough to overcome any preferential alignment in these materials. Due to the null resistance to spin realignment, the overall coercivity of these minute crystallites drops to zero. The

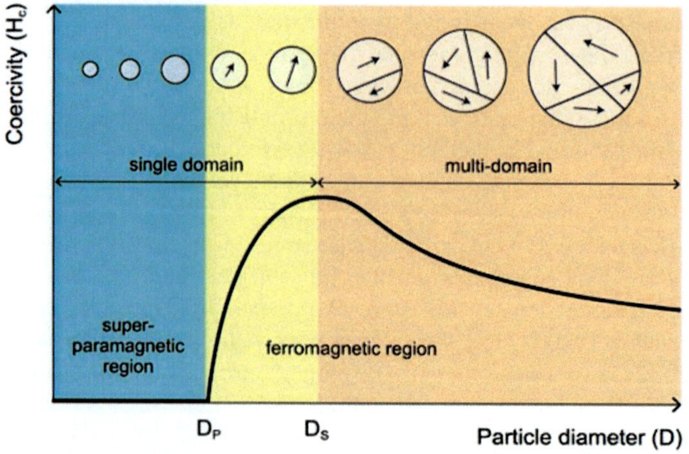

Fig. 3.2 Dependence of coercivity on the diameter of the magnetic particles

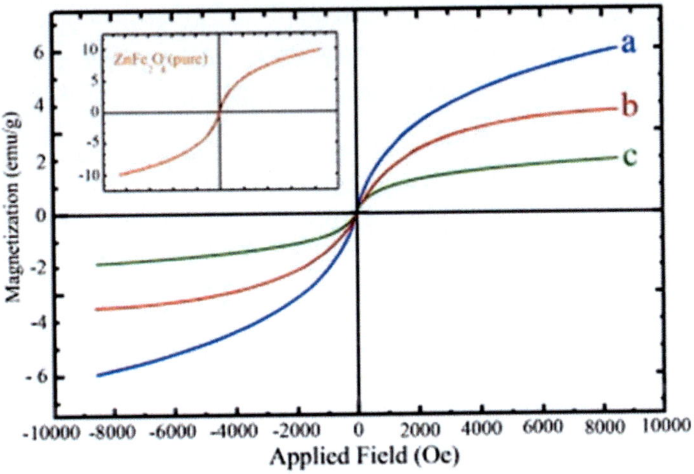

Fig. 3.3 M-H curves of superparamagnetic $ZnFe_2O_4$/graphene composites. Samples contain (a) 15, (b) 25, and (c) 40 wt% graphene. The inset shows the magnetic behavior of the pure zinc ferrite particles (Adapted from [11] with permission from Elsevier)

hysteresis loop vanishes, and thus remanence also goes to zero. An example of an M-H curve of an SPM ferrite can be seen in Fig. 3.3, which shows the M-H curves of $ZnFe_2O_4$/graphene composites with different weight fractions of the carbonaceous material. The curves show decreasing saturation with an increase in graphene content. Since this component does not contribute significantly to the magnetic character when compared to zinc ferrite, the specific magnetization of the samples (as a function of total mass) decreases accordingly.

The superparamagnetic character of ferrite nanoparticles may be tailored for a wide variety of applications. The lack of hysteresis losses has shown to be very useful in several fields, from chemical engineering to biomedicine. Perhaps the latter is the research field in which these particles show particular promise, especially as hyperthermia seeds.

Hyperthermia is a cancer treatment in which specific tissues are heated up, usually in a range between 41 and 46 °C. Given their increased metabolic rates, tumor cells are more sensitive to temperature than healthy ones. Therefore, localized heating is a promising, less invasive alternative to the current options such as surgery and chemotherapy, which have their major drawbacks. The cells can be locally heated via the incidence of radiation, such as infrared or microwaves. However, to reach the desired location, electromagnetic waves may also damage healthy cells on their way to the tumor.

The utilization of superparamagnetic particles circumvents the need for an external heat source. When injected into the patient's bloodstream, these particles can be directed toward specific tissues via the application of an external magnetic field, which works toward concentrating the particles in the desired tissue. When in place, an AC magnetic field causes the fast reversion of the magnetization direction of the particles. If the AC field's frequency is tailored to reverse faster than the relaxation of the particles, heat is dissipated locally due to the delay in magnetic relaxation [12]. This mechanism causes localized heat generation, minimizing thermal damage to nearby healthy tissue.

As should be evident, the materials used in hyperthermia treatment should be the least cytotoxic possible, as they are to be injected into the bloodstream of an already debilitated patient. Ferrite nanoparticles are usually very biocompatible and are quickly cleaned out of the body. Therefore, a large body of research on the utilization of spinel ferrites as hyperthermia seeds can be found in the literature [13–16]. The main property of interest regarding this biomedical application is the specific absorption rate (SAR) of the materials. This variable is defined as the power dissipated per mass of material and takes units of [W/g]. Ferrites usually display SAR values on the order of 10^2 W/g [17–19].

Given the different mechanisms responsible for power dissipation in an alternating magnetic field, morphological and chemical properties are critical in the definition of the SAR. Crystallite size plays a significant role in defining heat generation, but the chemical composition also shows a strong influence. Apostolov et al. [20] have conducted a detailed theoretical analysis on the effects of different divalent cations on the SAR of nanoferrites. The researchers have shown the existence of maxima in the SAR curves as a function of both particle size and composition of the magnetic nanoparticles (Fig. 3.4). Their work is a nuanced insight into the tailoring of these materials for their future application. The careful choice of particle size and divalent cation could lead to increased SAR, which would result in increased local heating. Conversely, the findings could hint at lower therapeutic dosages of the ferrites, decreasing the possibility of harmful side effects.

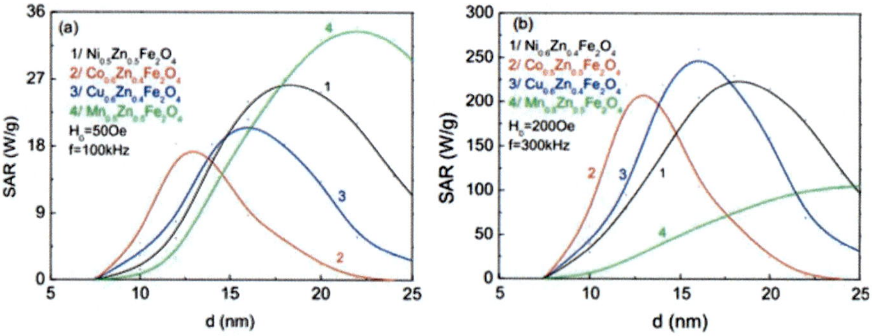

Fig. 3.4 SAR as a function of particle size for different zinc-substituted ferrites at alternating fields of 100 and 300 kHz (Adapted from [20] with permission from Springer)

3.3 Electrical Behavior

Ferrites are widely applied in the industry owing mostly to their electrical characteristic. They are one of a few classes of materials to display both a magnetic character and a high resistivity. Due to the simultaneous presence of these two characters, they are widely applied as the magnetic cores of RF transformers since the low conductivity helps prevent eddy currents. Nevertheless, not all ferrites display high resistivity. Indeed, these spinels show a wide range of electrical behaviors. Temperature and structural configuration are generally regarded as influential factors regarding the macroscopic electrical character.

As J.B. Goodenough explained in his seminal works [21–23], electron transfer in oxides occurs through two main pathways: cation-cation (c–c) and cation–anion–cation (c–a–c) interactions. These two mechanisms denote the possible overlap of d-orbitals of octahedral cations, which would allow for electrons to travel through the material. There is no possibility of overlap between d-orbitals of neighboring cations in corner-sharing octahedra, as the oxygen atom sits in between them. The exchange must therefore be mediated by the anion, and the mechanism is called c–a–c. In the case of spinel ferrites, the octahedral sites share an edge. In this arrangement, some electronic configurations allow for an effective overlap of the cationic d-orbital, creating a pathway along which the electrons can travel freely. Contrary to what takes place in c–a–c interactions, the charge carriers are delocalized. A similar effect is observed in face-sharing octahedra. Selected electron counting rules laid down by Goodenough can be seen in Table 3.1; the complete set of rules can be found in [21]. The exchange constant $\boldsymbol{J}$ denotes the expected coupling. The temperature $\boldsymbol{T}_{C,N}$ is estimated from oxides of transition elements of the first long period.

In practice, both c–c and c–a–c interactions are present in all materials. The relative strength of these two will determine whether an oxide will present metallic or semiconductive/insulating behavior. The first will be present when there is sufficient d-orbital overlap throughout the entire crystal (c–c), while the latter occurs when the electron transfer must be mediated by the anion. Since metallic conduction requires good overlap, it is also strongly dependent on the distance between the cations. Increases in distance between the octahedral sites, either from structural constraints or thermal dilation, exponentially affect the resistivity of the material.

In the case of spinel ferrites, their structural flexibility makes matters more complicated. The octahedral sites are not composed entirely of a single cation species but of

Table 3.1 Selected electron counting rules for edge-sharing octahedra based on Goodenough [21] (m_1, m_2—outer d-shell electrons; J—exchange constant; $T_{C,N}$—estimated temperature above which the effect disappears)

m_1	m_2	J	$T_{C,N}$ (K)
0	≤ 10	0	0
≤ 5	≤ 5	–	< 500
$5 < m < 8$	< 8	+	< 50
$8 \leq m \leq 10$	≤ 10	0	0

a complex mixture of different cations with different oxidation numbers, as denoted by their inversion degree (See Sect. 2.2). To further complicate matters, this combination of cations is not necessarily evenly distributed through the octahedral sites. Entropy considerations would dictate that the cations are arranged in the most random way possible. Therefore, even when some substantial d-overlap arises in the structure, it is not continuous enough for conduction to be effective. The visible result of this metallic jumble is that spinel ferrites usually (but not always!) display a semiconducting behavior at room temperature.

3.3.1 Semiconducting Mechanisms

Given the high localization of the electrons at the cationic sites and the subsequent little orbital overlap, band models are not the most appropriate tools when analyzing the conduction of ferrites. Localized electrons have very low mobility, which results in the high resistivity of spinel ferrites. In these materials, electron transfer is usually believed to occur via hopping between Fe^{2+} and Fe^{3+} pairs [24]. Nevertheless, a variety of mechanisms has been proposed to explain the precise mechanism of electronic conduction in ferrites. Correlated barrier hopping (CBH), overlapping large polaron tunneling (OLPT), non-overlapping small polaron tunneling (NSPT), and quantum mechanical tunneling (QMT) are some of the phenomena usually assigned to the driving force behind the charge mobility in ferrites.

When trying to pinpoint which of these pathways is predominant, AC impedance spectroscopy is a valuable tool. The spectra can be analyzed via Jonscher's power law [25]:

$$\sigma_{ac}(\omega) = \sigma_{dc} + A\omega^{s}$$

where σ_{ac} is the ac conductivity, σ_{dc} is the dc conductivity, A is a pre-exponential constant, ω is the frequency, and s is the power parameter. The latter denotes the dependence of AC conductivity on frequency. Thus, this power parameter gives great insight into the interaction of the spinel lattice with the charge carriers. The variation of this parameter with temperature is usually a good indication of which of the previously cited mechanisms is the main force behind charge transport.

If the power parameter is independent of temperature, quantum mechanical tunneling is the predominant mechanism. Radoń et al. [26] have shown this behavior in high entropy ferrites (HEF). The authors prepared (Zn,Mg,Ni,Fe,Cd)Fe_2O_4 via the co-precipitation of sulfate and nitrate precursors. The impedance spectra and power parameter of the product can be seen in Fig. 3.5. The s parameter obtained from Jonscher's power law shows almost no variation with temperature, which implies the QMT mechanism. The random cation distribution in these ferrites was pointed by the authors as a likely cause for this transport phenomenon since electron hopping would be practically negligible in the studied material due to the considerable distance

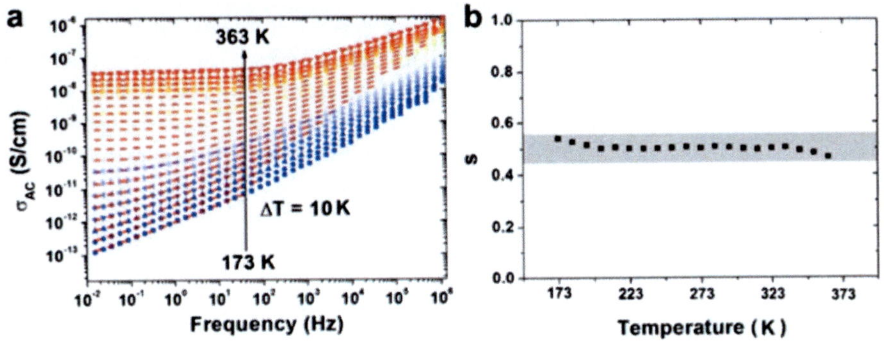

Fig. 3.5 AC electrical conductivity of (Zn,Mg,Ni,Fe,Cd)Fe_2O_4 high entropy ferrite, fit according to Jonscher's power law, (b) power parameter as a function of temperature (Adapted from [26] licensed under CC BY 2.0)

between Fe ions. The only possible pathway for the electrons involves tunneling from one localized state to the next.

A decrease of the power parameter with increasing temperature is an excellent indication that correlated barrier hopping (CBH) is the driving force behind charge transport. This model suggests charge transfer occurs due to thermal activation over the barrier between neighboring potential wells [27]. The power parameter can be defined as

$$s = 1 - \frac{6k_B T}{W_M + k_b T ln(1/\omega\tau_0)}$$

where k_B is the Boltzmann constant, T is the temperature, W_M is the polaron binding energy, and τ_0 is the relaxation time. Mechanisms defined by this equation are fairly common among spinel ferrites. Abdullah Dar et al. [28] have studied the conductivity of $Li_{0.35-0.5x}Cd_{0.3}Ni_xFe_{2.35-0.5x}O_4$, with x ranging from 0 to 0.08. The materials were prepared via a standard citrate–nitrate sol–gel method and treated at 700 °C for 2 h. Results have shown the predominance of the CHB conduction in a temperature range from 353 to 560 K. This mechanism is also observed in less complex ferrites. Nanoparticles of $CoFe_2O_4$ prepared via precipitation of sulfates have been shown to behave similarly [29]. Crystallites measuring approximately 12 nm show a clear negative slope when the power parameter is plotted as a function of temperature in a range between 323 and 573 K. Studies on $CoFe_2O_4$-polyvynilalcohol films show that this is the main pathway behind conductivity for ferrite loadings of up to 10 wt% [30]. Furthermore, the authors have demonstrated that the polaron binding energy W_M increases with higher ferrite contents, from 0.408 to 0.434 eV, as did the characteristic relaxation time τ_0 (1.09 to 9.81 ps).

A third conduction mechanism reveals itself when the power parameter increases with temperature: non-overlapping small polaron tunneling (NSPT), also known as small polaron tunneling. In this mechanism, the electron tunnels between states close to the Fermi level. However, the arrival of the electron relaxes the lattice around it, lowering the energy of this state. The relaxed lattice and the charge associated to it are known as a polaron. More specifically, in the case of NSPT, the lattice distortions associated with the electron do not overlap, so the activation energy necessary for charge transfer is not affected by the distance between neighboring states. The mathematical description of the power parameter in this case is written as:

$$s = 1 - \frac{4}{ln\left(\frac{1}{\omega\tau_0}\right) - \frac{W_H}{K_B T}}$$

where W_H is the hopping energy [31]. This type of mechanism is less concretely reported for ferrites. A study on $Ni_{0.6-x}Cd_xMg_{0.4}Fe_2O_4$ has shown an ingenious manner to prove the existence of NSPT mechanisms in ferrites [32]. The authors analyzed the impedance data via a double power law, i.e., a version of Jonscher's law with two frequency-dependent components. This model yielded two different power parameters. The s value related to the middle frequency range shows a behavior fitting of an NSPT mechanism.

The last mechanism is perhaps the one most commonly reported for spinel ferrites. The power parameter may yield a convex curve when plotted against temperature. The parameter s initially decreases with temperature increments, reaches a minimum, and subsequently increases. This behavior is characteristic of the overlapping large polaron tunneling (OLPT) mechanism. As proposed by Long [31], this phenomenon is very similar to the NPST mode, except that in this case the polarons are close enough to overlap. The interaction decreases the energetic barrier between the two localized states. The power parameter under such conditions can be described as:

$$s = 1 - \frac{8\alpha R_w + 6W_{H0}r_p/R_w k_B T}{\left(2\alpha R_w + W_{H0}r_p/R_w k_B T\right)^2}$$

where α is the spatial extent of the polaron, R_w is the tunneling distance, W_{H0} is associated with the activation energy of the transport, and r_p is the radius of the polaron. This equation leads to a power parameter with a minimum when plotted against temperature. An example of this behavior is shown in Fig. 3.6, which exhibits the thermal behavior of the power parameter of $CoFe_2O_4$ nanoparticles.

Veena Gopalan et al. [34] have shown that the OLPT mechanism is clearly predominant in nanostructured manganese ferrites prepared via co-precipitation. Studies on $Al_{0.2}Cd_{0.8}Fe_2O_4$ also give clear evidence that overlapping polarons are the main driving force behind conductivity in these more complex ferrites [35]. The authors show that the OLPT model explains the hopping of charge carriers between Fe^{2+} and Fe^{3+} between 293 and 613 K in this material. Kuru et al. [36] have discussed the effect of calcium-doping on the electrical properties of mixed zinc-nickel ferrites. The authors have linked the presence of calcium to the appearance of minima in the *s versus T* plots of these materials, which indicates the large polaron mechanism.

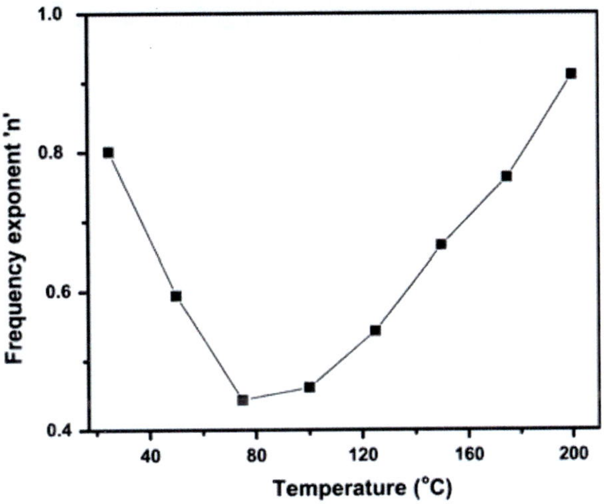

Fig. 3.6 Frequency exponent (or power parameter) of $CoFe_2O_4$ nanoparticles prepared via an auto-combustion method (Adapted from [33] with permission from Elsevier)

However, the inexistence of minima does not rule out the OLPT mechanism, as this clear indication might be lying outside the studied temperature range.

Different conduction mechanisms may also be present at different temperatures for the same ferrite. Radoń et al. [37] have shown a transition between two conduction models for magnetite nanoparticles prepared via co-precipitation using polyvinylpyrrolidone (PVP) as a surfactant. The authors assign the conductivity to a CBH model at low temperatures (below 260 K) and NSPT at higher thermal ranges (above 330 K). A smooth transition was found at intermediate temperatures, where the two mechanisms coexist. This study also goes further into the role of grains and grain boundaries in the overall conductivity of this ferrite. Selected results thereof can be seen in Fig. 3.7.

Despite being usually written off as semiconductors, spinel ferrites may also display metallic conduction under exceptional circumstances. For instance, some studies on nanostructured cobalt ferrite have detected a metallic-to-insulator transition close to room temperature. Rahman et al. [38] have shown that this material shows decreasing impedance under heating from 300 to 330 K. This phenomenon strongly implies semiconducting/insulating behavior. Above this temperature threshold, however, impedance increases with temperature, which is characteristic of metallic conduction. Delocalized electrons (the main charge carriers in metallic materials) have a decreased mean free path at higher temperatures. Thus, the resistivity/impedance of the material increases. The authors assigned this behavior to the mixed inversion of their ferrite, with cobalt ions almost equally distributed between tetrahedral and octahedral sites.

Venturini et al. [39] have also shown similar behavior in $CoFe_2O_4$ prepared via a sol–gel method and treated at 300 °C. The authors reported that the impedance (shown in Fig. 3.8) of pressed pellets of their material initially increased under heating up to 100 °C. Above this thermal threshold, the material showed a semiconducting behavior. The authors also assign the metallic character at the lower thermal range

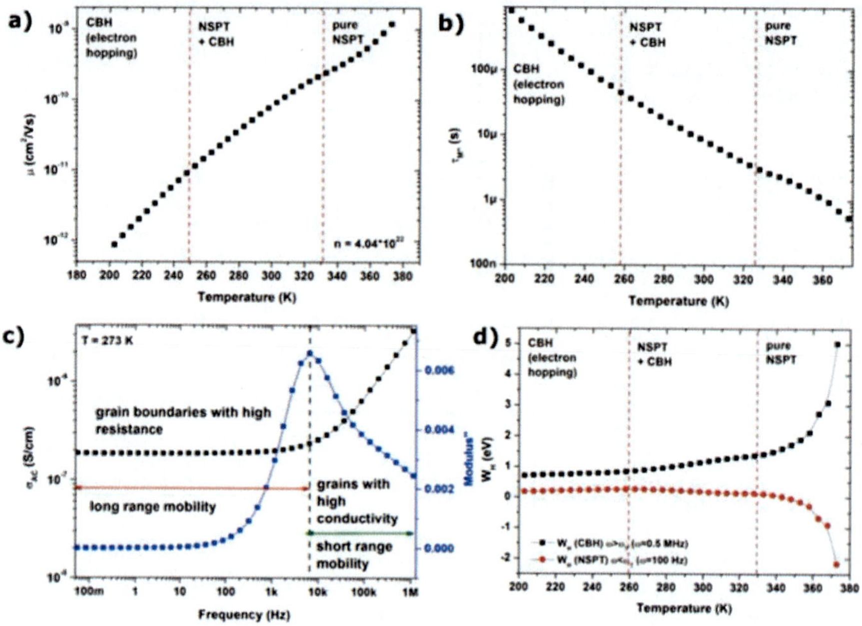

Fig. 3.7 **a** Electron mobility and **b** relaxation times associated with conduction in Fe_3O_4. **c** Analysis of dielectric properties and AC conductivity and **d** value of hopping energy for CBH and NSPT models at 100 Hz and 500 kHz (Adapted from [37] licensed with CC BY 4.0)

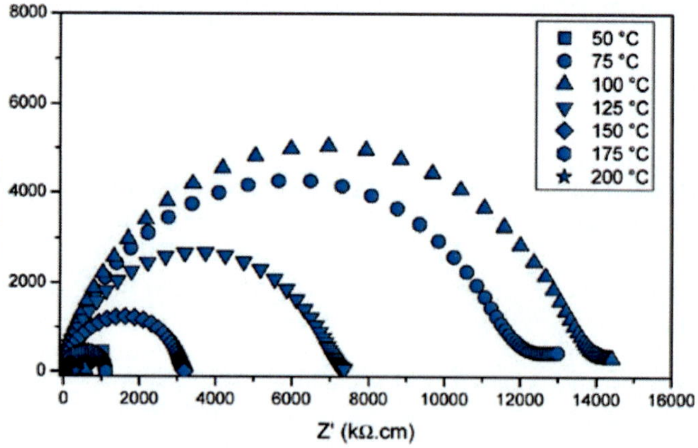

Fig. 3.8 Impedance spectra of $CoFe_2O_4$ nanoparticles prepared via a citrate/nitrate sol–gel method (Adapted from Venturini et al. [39] with permission from The Royal Society of Chemistry

to the cationic arrangement in the ferrite, which was analyzed via Mössbauer spectroscopy. The cationic placement favored metallic conduction, as defined by Goodenough (Table 3.1). Furthermore, the increase in temperature leads to the expansion of the unit cell, which also works toward decreasing the orbital overlap necessary for electron delocalization. An endothermic event was also observed above 50 °C in the DSC curves of the material, which might indicate a charge-ordering transition.

Metallic-to-insulator transitions (MIT) are also very commonly observed in perovskite ferrites. Several mechanisms are used to explain this phenomenon. In $BiFeO_3$ films, for instance, this transition may be explained by a crystallographic transformation [40]. The transition from the orthorhombic β phase to the cubic γ polymorph causes the band gap to collapse toward zero. Zhang et al. [41] have modeled the electrical behavior of alkaline earth ferrites in an attempt to elucidate the reasons for the presence or absence of MIT in this family of compounds. The authors clarify why strontium and barium ferrites remain metallic down to low temperatures, while $CaFeO_3$ undergoes a semiconductor transition. The work assigns the transition to the rotation of motions of the oxygen anions and points out mechanisms by which an MIT could be tailored into the structure of perovskite ferrites.

Several such metallic-to-insulator transitions have been described in the literature for ferrites, but one particular phenomenon has come to be inextricably associated with ferrites: the Verwey transition.

3.3.2 *Verwey Transition*

In 1939, Verwey reported his eponymous transition when studying the electrical behavior of magnetite (Fe_3O_4) at low. temperatures [42]. Below 120 K, this material showed a sudden increase in conductivity. An example of this drastic effect can be seen in Fig. 3.9. Anomalies in lattice constants and specific heat were also observed

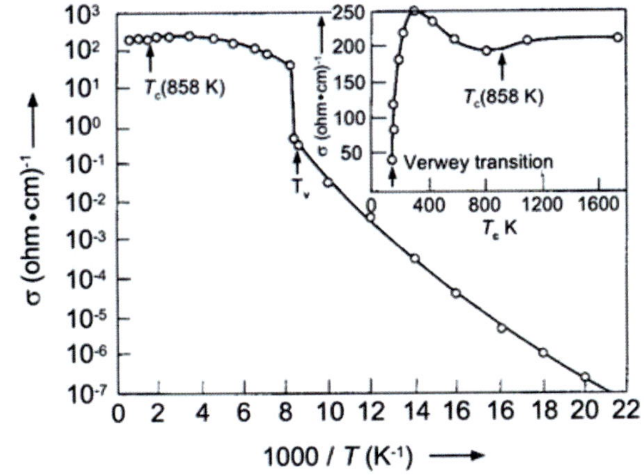

Fig. 3.9 Thermal behavior of the conductivity of a magnetite single crystal (Adapted from [48] with permission from the American Physics Society)

at 114 K. Verwey proposed that the decrease in temperature led to an ordering of Fe^{2+} and Fe^{3+} ions in the magnetite lattice, which should explain the observed phenomena. At low temperatures, the charge hopping between Fe^{2+} and Fe^{3+} should cease, and the cations should arrange themselves in a regular pattern [43]. Since charge-ordering (CO) transformations are fundamental to the application of several other classes of materials [44–46], Verwey transitions (VT) have been closely studied ever since. Nevertheless, the principles behind this phenomenon in spinel ferrites were and are still hotly debated. F. Walz wrote at length about the historical aspects and advances in this discussion [47].

Initially, the difficulty in producing high-quality, single-crystal magnetite led to the proposal of several misleading theories to explain this transition. Verwey himself showed in his seminal work the marked effect of oxygen vacancies on the onset of the transition [42]. An excess of Fe_2O_3 strongly suppressed the drastic drop in conductivity. From very early on, scientists observed that the stoichiometry and the structure of the studied ferrite played a major role in the presence or absence of the VT.

Liu et al. [49] have studied the effects of oxygen stoichiometry and thickness on the VT of Fe_3O_4 thin films. The authors have shown that the oxygen partial pressure in the film growth chamber has a strong effect on the onset of the transition. Even more dramatic is the importance of film thickness. Decreasing thicknesses produced an ever-smoother resistivity transition. At 5 nm, the phenomenon was wholly suppressed. Size effects have also been analyzed for Fe_3O_4 nanoparticles. Lee et al. [50] studied the effect of particle size on the resistivity change. The authors observed that the onset temperature of VT shifts toward lower values with decreasing particle sizes. This effect becomes more pronounced for crystallites smaller than 20 nm. The transition was suppressed entirely for particles with a radius of 6 nm.

Though ingenious, the charge ordering model proposed by Verwey has shown to be only marginally correct. Currently, the most accepted model to explain the charge ordering in these strongly correlated systems was postulated by Senn, Wright, and Attfield [51]. The authors have analyzed the Cc low-temperature phase of magnetite, observing the shortening of some Fe–Fe bonds. The findings suggest that the localized electrons are distributed along so-called trimerons, proposed quasiparticles that are formed of linear three-Fe-site units. A detailed image of the trimeron structure in a magnetite crystal is shown in Fig. 3.10. The charge order and site displacement should explain the electrical and magnetic anomalies observed in the Verwey transition.

Other than magnetite, several spinel ferrites also show transitions consistent with charge-ordering transformations. Research has shown that MFe_2O_4 thin films (M = Zn, Co, Ni) also display a resistivity drop when heated above approximately 100 K [53]. Datt et al. [54] have observed magnetic anomalies in solvothermal nickel–cobalt ferrites in temperatures that would indicate a Verwey-type transition. The cationic disorder and spherical shape of the as-synthesized particles suppressed possible transitions. However, the particles become octahedral upon annealing. The authors report that the thermal treatment (and its consequences on the size and cationic distribution) led to the appearance of a magnetic anomaly at 151 K, which was assigned to a charge-ordering transformation.

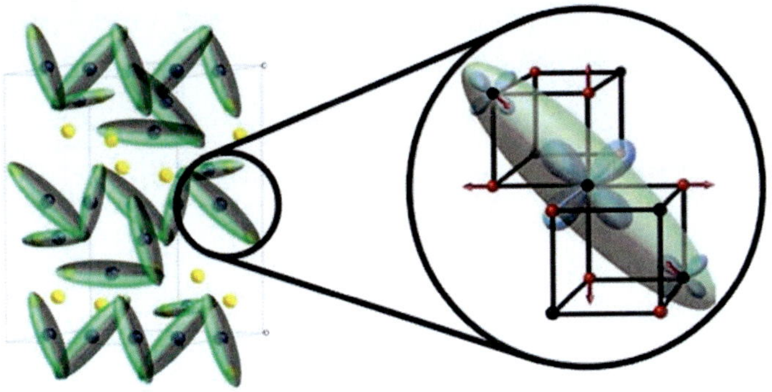

Fig. 3.10 Charge-ordered structure of magnetite, showing the trimeron organization. Blue and yellow spheres show Fe^{2+} and Fe^{3+}, respectively. The inset shows the structure of a single trimeron (Adapted from [51, 52] with permission from Nature)

3.4 Optical Properties

Although their most common applications involve their electrically insulating character, the band structure of spinel ferrites undoubtedly classifies them as actual semiconductors. Their band gap usually falls in a range between 0.8 and 2.0 V—the insulating behavior comes mostly from the minimal charge carrier transport in the spinel lattice, as discussed in the previous section. The low band gap energies typically render a spinel ferrite black or, at most, a dark reddish-brown color.

Chavarriaga et al. have studied the optical properties of several ferrites prepared by gel combustion synthesis using 6-aminohexanoic acid as fuel [55]. The reflectance spectra of $CoFe_2O_4$, $MgFe_2O_4$, and $ZnFe_2O_4$ obtained via this method are shown in Fig. 3.11. All three ferrites show low reflectance in the high-energy end of the visible spectrum, indicating a strong absorption of these wavelengths. On the other hand,

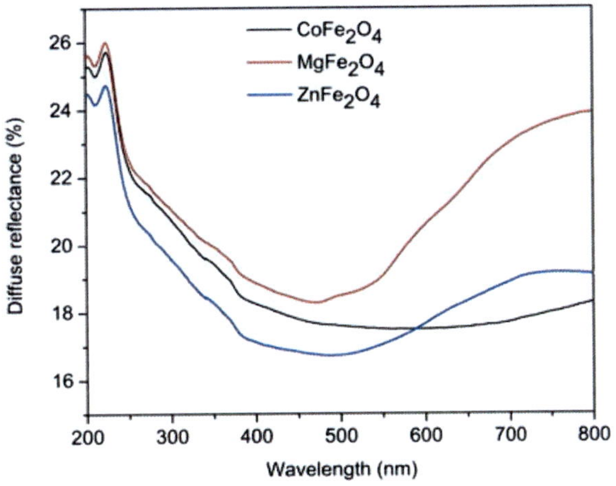

Fig. 3.11 Diffuse reflectance spectra of cobalt, magnesium, and zinc ferrites obtained via a gel combustion method (Adapted from [55] with permission from Elsevier)

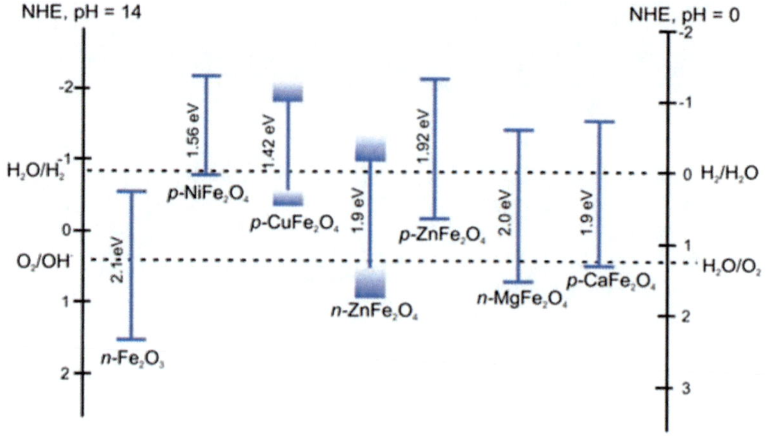

Fig. 3.12 Band positions of selected spinel ferrites in aqueous solutions versus NHE (Reprinted from [59] licensed under CC BY 4.0)

zinc and magnesium ferrites show some reflectance of low-energy visible photons, which explains the reddish hue of these ferrites.

Figure 3.12 shows the band positions of several spinel ferrites in aqueous electrolytes compared against an NHE reference. As might be evident for the reader at this point in the book, the band gap of spinel ferrites is not as well-behaved as this figure might make it seem. For instance, the literature reports band gap energies between 0.1 and 2.6 eV for $CoFe_2O_4$ [56–58]. Several explanations for these widely varying measurements in similar materials have been proposed.

One hypothesis for the band gap variation of spinel ferrites is the influence of oxygen vacancies. Erhardt and collaborators [57] studied the effects of sucrose on the optical gap of cobalt ferrites. Samples were prepared via a modified sol–gel method, with the addition of different amounts of the carbohydrate as an excess reducing agent. The authors have found that the presence of the additive leads to marked changes in the structural and magnetic properties of the products. Increasing amounts of sucrose lead to an ever more reducing reactional environment, which should lead to the creation of oxygen defects in the spinel lattice. Consequently, the observed band gap shows a linear inverse relation with the concentration of sucrose in the reaction in the range between 0 and 300 mmol.L^{-1}; the results are reproduced in Fig. 3.13. The authors also raise the possibility of another factor at play in defining the band gap of these ferrites: the inversion degree.

The effect of the inversion degree on the band structure of ferrites was shown in an exciting work by Soliman and collaborators [60]. The authors have used local density approximation to describe the electronic and magnetic properties of a normal spinel $ZnFe_2O_4$, an antiferromagnetic insulator. Previous simulations showed metallic behavior for this material, and the authors have looked into addressing this issue. Metallic conduction is expected of normal spinels containing complete d or

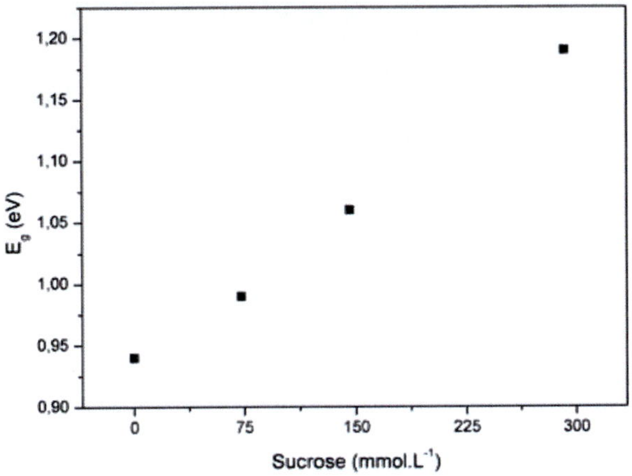

Fig. 3.13 Band gap of $CoFe_2O_4$ as a function of the concentration of sucrose in the reactional medium (Reprinted from [57] with permission from Elsevier)

f shells, as is the case of the studied compound. The authors show that the application of Hubbard parameters is not sufficient to describe zinc ferrites fully. The study concludes that $ZnFe_2O_4$ cannot be treated as a correlated system and that partial inversion and site deformation need to be taken into account.

Another article that discusses at length the effect of inversion degree and oxygen vacancies on the optical band gap of ferrites was published by Dileep and coworkers [61]. The authors presented the results of DFT calculations of several crystalline configurations of $NiFe_2O_4$ and $CoFe_2O_4$ in an attempt to relate the experimentally measured band gap values of these materials to the fundamental phenomena affecting this property. The effect of two common structural parameters, the inversion degree and cation vacancies, is clearly shown to have a profound influence on the observed band gap. For instance, the calculated band gaps for different $NiFe_2O_4$ species are reproduced in Fig. 3.14. Different inversion degrees markedly affect the band struc-

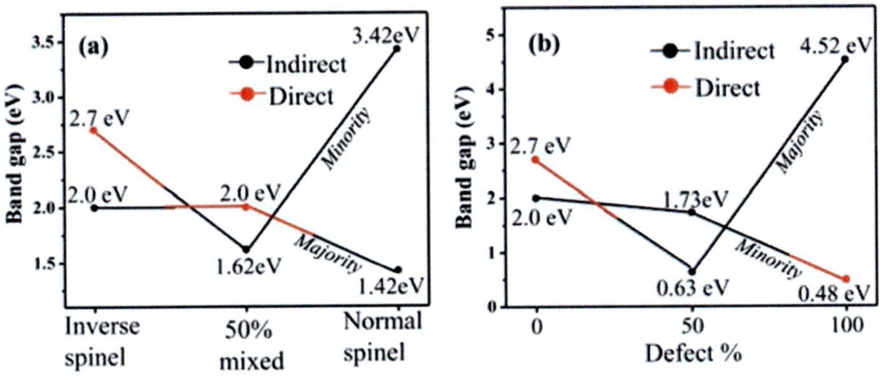

Fig. 3.14 Band gap of different configurations of $NiFe_2O_4$. On the left side, the effect of varying inversion degrees; on the right, the effect of tetrahedral cation vacancies (Reproduced from [61] with permission from the American Institute of Physics)

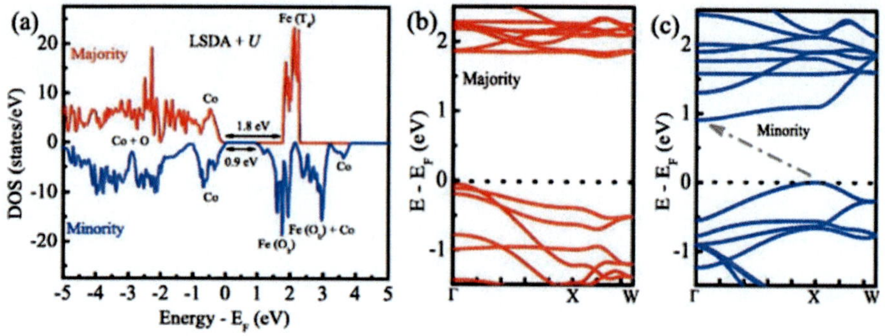

Fig. 3.15 **a** Density of states of $CoFe_2O_4$ and energy bands along G-X-W for the **b** majority and **c** minority channels (Reproduced from [62] with permission from the American Institute of Physics)

ture of nickel ferrites, and this effect varies for the minority and majority channels (opposite spin orientations). Furthermore, the pathway of these band gaps also switches from direct to indirect and vice versa. The authors also show the effect of a common phenomenon seen in spinel ferrites, cationic vacancies, a type of structural relaxation that was previously discussed in Chapter 2. The study exaggerated this feature to 50 and 100% tetrahedral vacancies to demonstrate the influence of this parameter. The results show an effect of similar magnitude to that of the inversion degree. Structural vacancies may both increase and decrease the calculated band gap, with different consequences for each of the channels.

Holinsworth and collaborators have conducted a study with the same materials, wherein they determined whether the band gap of these ferrites was direct or indirect [62]. The calculated density of states (DOS) and selected energy bands for the majority and minority channels of $CoFe_2O_4$ are shown in Fig. 3.15. The results clearly show an indirect band gap along the minority channel for cobalt ferrite. The authors extracted a 1.2 eV indirect gap and a higher-energy 2.7 eV direct gap. Energies lower than those observed for $NiFe_2O_4$ are primarily assigned to ionic size effects; smaller nickel ions induce deeper states due to more intense electrostatic interactions. The study also reports a further relevant result. Their calculations show that a partially inverted structure (75% inversion) has higher energy than the fully inverted configuration, implying that the mixed arrangement might be a metastable state for $CoFe_2O_4$.

The intense light absorption of ferrites, coupled to their suitable bandgap, has led to intense research on the application of this family of materials as photoactive electrodes in water-splitting reactors. Spinel ferrites are photochemically stable and absorb visible light, both of which are requirements for application as photoelectrodes. Their structural versatility also means their conduction and valence bands may be tuned to better fit the desired electrochemical system. In particular, $ZnFe_2O_4$ has been intensively studied for application as a photoanode in water-splitting cells [63].

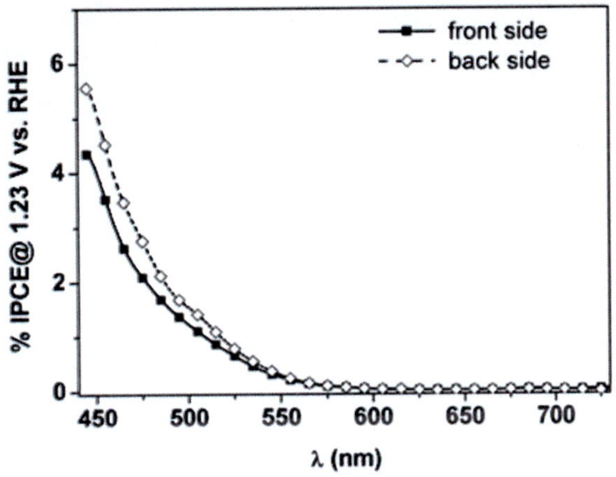

Fig. 3.16 IPCE spectra of $ZnFe_2O_4$ films illuminated from the front and back sides (Reprinted with permission from [64]. Copyright (2017) American Chemical Society)

One of the properties of interest in photoelectrodes is their IPCE (incident photon-to-electron conversion efficiency), which is a measure of their quantum efficiency. This variable indicates the fraction of the light quanta that is effectively converted by the material into mobile electrons. An ideal IPCE value of 100% would mean that every incident photon was effectively transformed into a free charge, with no energy loss via secondary mechanisms. In natural materials, this conversion efficiency is dependent on the wavelength of the photons and the intrinsic properties of the analyzed material.

Peeters et al. [64] have studied the photoactivity of zinc ferrite films. The authors deposited this material via a solvent-free CVD approach at temperatures between 500 and 800 °C. The IPCE spectra of the produced materials can be seen in Fig. 3.16. The authors have shown a conversion efficiency of 5.2% for photons with a wavelength of 450 nm for $ZnFe_2O_4$ films with a thickness of 340 nm. The different IPCE values when illuminated from opposite sides are seen as evidence of porosity in the thick films. Guijarro and collaborators have shown similar IPCE values for high-energy visible photons in several spinel ferrites [65]. The authors report efficiencies of 4% at 400 nm for cobalt and magnesium ferrites coated with nickel ferrite. An IPCE of 8% was reached at similar conditions for $ZnFe_2O_4$ annealed under hydrogen and coated with $NiFe_2O_4$.

References

1. Zabotto FL, Gualdi AJ, Eiras JA, de Oliveira AJA, Garcia D (2012) Influence of the sintering temperature on the magnetic and electric properties of $NiFe_2O_4$ ferrites. Mater Res 15:428–433
2. Chauhan L, Shukla AK, Sreenivas K (2015) Dielectric and magnetic properties of Nickel ferrite ceramics using crystalline powders derived from DL alanine fuel in sol–gel auto-combustion. Ceram Int 41:8341–8351. https://doi.org/10.1016/J.CERAMINT.2015.03.014

3. Azizi A, Sadrnezhaad SK (2010) Effects of annealing on phase evolution, microstructure and magnetic properties of mechanically synthesized nickel-ferrite. Ceram Int 36:2241–2245. https://doi.org/10.1016/J.CERAMINT.2010.06.004
4. Dho J, Lee EK, Park JY, Hur NH (2005) Effects of the grain boundary on the coercivity of barium ferrite $BaFe_{12}O_{19}$. J Magn Magn Mater 285:164–168. https://doi.org/10.1016/J.JMMM.2004.07.033
5. Liu X, Wang J, Gan L-M, Ng S-C, Ding J (1998) An ultrafine barium ferrite powder of high coercivity from water-in-oil microemulsion. J Magn Magn Mater 184:344–354. https://doi.org/10.1016/S0304-8853(97)01141-4
6. Wang S, Ding J, Shi Y, Chen Y (2000) High coercivity in mechanically alloyed $BaFe_{10}A_{12}O_{19}$. J Magn Magn Mater 219:206–212. https://doi.org/10.1016/S0304-8853(00)00450-9
7. Wang HZ, Hai YN, Yao B, Xu Y, Shan L, Xu L, Tang JL, Wang QH (2017) Tailoring structure and magnetic characteristics of strontium hexaferrite via Al doping engineering. J Magn Magn Mater 422:204–208. https://doi.org/10.1016/J.JMMM.2016.08.066
8. Houshiar M, Zebhi F, Razi ZJ, Alidoust A, Askari Z (2014) Synthesis of cobalt ferrite ($CoFe_2O_4$) nanoparticles using combustion, coprecipitation, and precipitation methods: a comparison study of size, structural, and magnetic properties. J Magn Magn Mater 371:43–48. https://doi.org/10.1016/J.JMMM.2014.06.059
9. Cedeño-Mattei Y, Perales-Pérez O, Uwakweh ONC (2012) Synthesis of high-coercivity non-stoichiometric cobalt ferrite nanocrystals: structural and magnetic characterization. Mater Chem Phys 132:999–1006. https://doi.org/10.1016/J.MATCHEMPHYS.2011.12.049
10. Kambale RC, Shaikh PA, Kamble SS, Kolekar YD (2009) Effect of cobalt substitution on structural, magnetic and electric properties of nickel ferrite. J Alloys Compd 478:599–603. https://doi.org/10.1016/J.JALLCOM.2008.11.101
11. Meidanchi A, Akhavan O (2014) Superparamagnetic zinc ferrite spinel–graphene nanostructures for fast wastewater purification, Carbon 69:230–238. https://doi.org/10.1016/j.carbon.2013.12.019
12. Sharifi I, Shokrollahi H, Amiri S (2012) Ferrite-based magnetic nanofluids used in hyperthermia applications. J Magn Magn Mater 324:903–915. https://doi.org/10.1016/j.jmmm.2011.10.017
13. Fantechi E, Innocenti C, Zanardelli M, Fittipaldi M, Falvo E, Carbo M, Shullani V, Di Cesare Mannelli L, Ghelardini C, Ferretti AM, Ponti A, Sangregorio C, Ceci P (2014) A smart platform for hyperthermia application in Cancer treatment: cobalt-doped ferrite nanoparticles mineralized in human ferritin cages. ACS Nano 8:4705–4719. https://doi.org/10.1021/nn500454n.
14. He S, Zhang H, Liu Y, Sun F, Yu X, Li X, Zhang L, Wang L, Mao K, Wang G, Lin Y, Han Z, Sabirianov R, Zeng H (2018) Maximizing specific loss power for magnetic hyperthermia by hard-soft mixed ferrites. Small 14:1800135. https://doi.org/10.1002/smll.201800135
15. Mameli V, Musinu A, Ardu A, Ennas G, Peddis D, Niznansky D, Sangregorio C, Innocenti C, Thanh NTK, Cannas C (2016) Studying the effect of Zn-substitution on the magnetic and hyperthermic properties of cobalt ferrite nanoparticles. Nanoscale 8:10124–10137. https://doi.org/10.1039/C6NR01303A
16. Du Y, Liu X, Liang Q, Liang X-J, Tian J (2019) Optimization and design of magnetic ferrite nanoparticles with uniform Tumor distribution for highly sensitive MRI/MPI performance and improved magnetic hyperthermia therapy. Nano Lett 19:3618–3626. https://doi.org/10.1021/acs.nanolett.9b00630
17. Mazario E, Menéndez N, Herrasti P, Cañete M, Connord V, Carrey J (2013) Magnetic hyperthermia properties of electrosynthesized cobalt ferrite nanoparticles. J Phys Chem C 117:11405–11411. https://doi.org/10.1021/jp4023025
18. Doaga A, Cojocariu AM, Amin W, Heib F, Bender P, Hempelmann R, Caltun OF (2013) Synthesis and characterizations of manganese ferrites for hyperthermia applications. Mater Chem Phys 143:305–310. https://doi.org/10.1016/j.matchemphys.2013.08.066
19. Céspedes E, Byrne JM, Farrow N, Moise S, Coker VS, Bencsik M, Lloyd JR, Telling ND (2014) Bacterially synthesized ferrite nanoparticles for magnetic hyperthermia applications. Nanoscale 6:12958–12970. https://doi.org/10.1039/C4NR03004D

20. Apostolov A, Apostolova I, Wesselinowa J (2019) Specific absorption rate in Zn-doted ferrites for self-controlled magnetic hyperthermia. Eur Phys J B 92:58. https://doi.org/10.1140/epjb/e2019-90567-2
21. Goodenough JB (1960) Direct cation- -cation interactions in several oxides. Phys Rev 117:1442–1451. https://doi.org/10.1103/PhysRev.117.1442
22. Goodenough JB (1958) An interpretation of the magnetic properties of the perovskite-type mixed crystals La1−xSrxCo$O_3-\lambda$. J Phys Chem Solids 6:287–297. https://doi.org/10.1016/0022-3697(58)90107-0
23. Goodenough JB (1955) Theory of the role of covalence in the Perovskite-type manganites [La, *M*(II)] Mn O_3. Phys Rev 100:564–573. https://doi.org/10.1103/PhysRev.100.564
24. Hosseinpour A, Sadeghi H, Morisako A (2007) Simulation of DC-hopping conduction in spinel ferrites using free electron gas model. J Magn Magn Mater 316:e283–e286. https://doi.org/10.1016/j.jmmm.2007.02.119
25. Jonscher AK (1977) The 'universal' dielectric response. Nature 267:673. https://doi.org/10.1038/267673a0
26. Radoń A, Hawełek Ł, Łukowiec D, Kubacki J, Włodarczyk P (2019) Dielectric and electromagnetic interference shielding properties of high entropy (Zn,Fe,Ni,Mg,Cd)Fe_2O_4 ferrite. Sci Rep 9:20078. https://doi.org/10.1038/s41598-019-56586-6
27. Megha R, Kotresh S, Ravikiran YT, Ramana CHVV, Vijaya Kumari SC, Thomas S (2017) Study of alternating current conduction mechanism in polypyrrole-magnesium ferrite hybrid nanocomposite through correlated barrier hopping model. Compos Interfaces 24:55–68. https://doi.org/10.1080/09276440.2016.1185298
28. Abdullah Dar M, Majid K, Batoo KM, Kotnala RK (2015) Dielectric and impedance study of polycrystalline $Li_{0.35-0.5X}Cd_{0.3}NiXFe_{2.35-0.5X}O_4$ ferrites synthesized via a citrate-gel auto combustion method. J Alloys Compd 632:307–320. https://doi.org/10.1016/j.jallcom.2015.01.190
29. Abu-Samaha FSH, Ismail MIM (2014) AC conductivity of nanoparticles $Co_xFe_{(1-x)}Fe_2O_4$ (x=0, 0.25 and 1) ferrites. Mater Sci Semicond Process 19:50–56. https://doi.org/10.1016/j.mssp.2013.11.027
30. Rana DK, Singh SK, Kundu SK, Choudhary RJ, Basu S (2018) Electrical and magnetic properties of polyvinyl alcohol–cobalt ferrite nanocomposite films. Bull Mater Sci 41:92. https://doi.org/10.1007/s12034-018-1608-0
31. Long AR (1982) Frequency-dependent loss in amorphous semiconductors. Adv Phys 31:553–637. https://doi.org/10.1080/00018738200101418
32. Hamdaoui N, Azizian-Kalandaragh Y, Khlifi M, Beji L (2019) Cd-doping effect on morphologic, structural, magnetic and electrical properties of $Ni_{0.6-x}Cd_xMg_{0.4}Fe_2O_4$ spinel ferrite ($0 \leq x \leq 0.4$). J Alloys Compd 803:964–970. https://doi.org/10.1016/j.jallcom.2019.06.339
33. Panda RK, Muduli R, Kar SK, Behera D (2014) Dielectric relaxation and conduction mechanism of cobalt ferrite nanoparticles. J Alloys Compd 615:899–905. https://doi.org/10.1016/j.jallcom.2014.07.031
34. Veena Gopalan E, Malini KA, Saravanan S, Sakthi Kumar D, Yoshida Y, Anantharaman MR (2008) Evidence for polaron conduction in nanostructured manganese ferrite. J Phys D Appl Phys 41:185005. https://doi.org/10.1088/0022-3727/41/18/185005
35. Kuru TŞ, Şentürk E, Eyüpoğlu V (2017) Overlapping large Polaron conductivity mechanism and dielectric properties of $Al_{0.2}Cd_{0.8}Fe_2O_4$ Ferrite Nanocomposite. J Supercond Nov Magn 30:647–655. https://doi.org/10.1007/s10948-016-3847-x
36. Kuru M, Şaşmaz Kuru T, Bağcı S (2019) The role of the calcium concentration effect on the structural and dielectric properties of mixed Ni–Zn ferrites. J Mater Sci Mater Electron 30:5438–5453. https://doi.org/10.1007/s10854-019-00837-9
37. Radoń A, Łukowiec D, Kremzer M, Mikuła J, Włodarczyk P (2018) Electrical conduction mechanism and dielectric properties of spherical shaped Fe_3O_4 nanoparticles synthesized by co-precipitation method. Mater (Basel, Switzerland) 11:735. https://doi.org/10.3390/ma11050735

38. ur Rahman A, Rafiq MA, Karim S, Maaz K, Siddique M, Hasan MM (2011) Semiconductor to metallic transition and polaron conduction in nanostructured cobalt ferrite. J Phys D Appl Phys 44:165404. https://doi.org/10.1088/0022-3727/44/16/165404
39. Venturini J, Zampiva RYS, Piva DH, Piva RH, da Cunha JBM, Bergmann CP (2018) Conductivity dynamics of metallic-to-insulator transition near room temperature in normal spinel $CoFe_2O_4$ nanoparticles. J Mater Chem C 6:4720–4726. https://doi.org/10.1039/C8TC00099A
40. Palai R, Katiyar RS, Schmid H, Tissot P, Clark SJ, Robertson J, Redfern SAT, Catalan G, Scott JF (2008) β phase and $\gamma-\beta$ metal-insulator transition in multiferroic $BiFeO_3$. Phys Rev B 77:14110. https://doi.org/10.1103/PhysRevB.77.014110
41. Zhang Y, Schmitt MM, Mercy A, Wang J, Ghosez P (2018) From charge- to orbital-ordered metal-insulator transition in alkaline-earth ferrites. Phys Rev B 98:81108. https://doi.org/10.1103/PhysRevB.98.081108
42. Verwey EJW (1939) Electronic conduction of magnetite (Fe_3O_4) and its transition point at low temperatures. Nature 144:327–328. https://doi.org/10.1038/144327b0
43. Bohra M, Agarwal N, Singh V (2019) A short review on verwey transition in nanostructured Fe_3O_4 Materials. J Nanomater 2019:8457383. https://doi.org/10.1155/2019/8457383
44. Comin R, Sutarto R, He F, da Silva Neto EH, Chauviere L, Fraño A, Liang R, Hardy WN, Bonn DA, Yoshida Y, Eisaki H, Achkar AJ, Hawthorn DG, Keimer B, Sawatzky GA, Damascelli A (2015) Symmetry of charge order in cuprates. Nat Mater 14:796–800. https://doi.org/10.1038/nmat4295
45. Hervieu M, Barnabé A, Martin C, Maignan A, Damay F, Raveau B (1999) Evolution of charge ordering in manganites. Eur Phys J B—Condens Matter Complex Syst 8:31–41. https://doi.org/10.1007/s100510050665
46. Adler P, Jeglič P, Reehuis M, Geiß M, Merz P, Knaflič T, Komelj M, Hoser A, Sans A, Janek J, Arčon D, Jansen M, Felser C (2018) Verwey-type charge ordering transition in an open-shell *p*-electron compound. Sci Adv 4:eaap7581. https://doi.org/10.1126/sciadv.aap7581.
47. Walz F (2002) The Verwey transition—a topical review. J Phys Condens Matter 14:R285–R340. https://doi.org/10.1088/0953-8984/14/12/203
48. Miles PA, Westphal WB, Von Hippel A (1957) Dielectric spectroscopy of ferromagnetic semiconductors. Rev Mod Phys 29:279–307. https://doi.org/10.1103/RevModPhys.29.279
49. Liu XH, Rata AD, Chang CF, Komarek AC, Tjeng LH (2014) Verwey transition in Fe_3O_4 thin films: Influence of oxygen stoichiometry and substrate-induced microstructure. Phys Rev B 90:125142. https://doi.org/10.1103/PhysRevB.90.125142
50. Lee J, Kwon SG, Park J-G, Hyeon T (2015) Size dependence of metal-insulator transition in stoichiometric Fe_3O_4 nanocrystals. Nano Lett 15:4337–4342. https://doi.org/10.1021/acs.nanolett.5b00331
51. Senn MS, Wright JP, Attfield JP (2012) Charge order and three-site distortions in the verwey structure of magnetite. Nature 481:173–176. https://doi.org/10.1038/nature10704
52. Attfield JP (2014) The verwey phase of magnetite: a long-running mystery in ferrites. 粉体および粉末冶金 61:S43–S48. https://doi.org/10.2497/jjspm.61.S43
53. Brachwitz K, Böntgen T, Lorenz M, Grundmann M (2013) On the transition point of thermally activated conduction of spinel-type MFe_2O_4 ferrite thin films (M = Zn Co, Ni). Appl Phys Lett 102:172104. https://doi.org/10.1063/1.4803475
54. Datt G, Sen Bishwas M, Manivel Raja M, Abhyankar AC (2016) Observation of magnetic anomalies in one-step solvothermally synthesized nickel–cobalt ferrite nanoparticles. Nanoscale 8:5200–5213. https://doi.org/10.1039/C5NR06791J
55. Chavarriaga EA, Lopera AA, Franco V, Bergmann CP, Alarcón J (2020) Gel combustion synthesis and magnetic properties of $CoFe_2O_4$, $ZnFe_2O_4$, and $MgFe_2O_4$ using 6-aminohexanoic acid as a new fuel. J Magn Magn Mater 497:166054. https://doi.org/10.1016/j.jmmm.2019.166054.
56. Waldron RD (1955) Infrared spectra of ferrites. Phys Rev 99:1727–1735. https://doi.org/10.1103/PhysRev.99.1727
57. Erhardt CS, Caldeira LE, Venturini J, Bragança SR, Bergmann CP (2020) Sucrose as a sol-gel synthesis additive for tuning spinel inversion and improving the magnetic properties of

$CoFe_2O_4$ nanoparticles. Ceram Int 46:12759–12766. https://doi.org/10.1016/j.ceramint.2020.02.044
58. Ravindra AV, Padhan P, Prellier W (2012) Electronic structure and optical band gap of $CoFe_2O_4$ thin films. Appl Phys Lett 101:161902. https://doi.org/10.1063/1.4759001
59. Taffa DH, Dillert R, Ulpe AC, Bauerfeind KCL, Bredow T, Bahnemann DW, Wark M (2016) Photoelectrochemical and theoretical investigations of spinel type ferrites ($M_xFe_{3-x}O_4$) for water splitting: a mini-review. J Photonics Energy 7:1–25. https://doi.org/10.1117/1.JPE.7.012009
60. Soliman S, Elfalaky A, Fecher GH, Felser C (2011) Electronic structure calculations for $ZnFe_2O_4$. Phys Rev B 83:85205. https://doi.org/10.1103/PhysRevB.83.085205
61. Dileep K, Loukya B, Pachauri N, Gupta A, Datta R (2014) Probing optical band gaps at the nanoscale in $NiFe_2O_4$ and $CoFe_2O_4$ epitaxial films by high resolution electron energy loss spectroscopy. J Appl Phys 116:103505. https://doi.org/10.1063/1.4895059
62. Holinsworth BS, Mazumdar D, Sims H, Sun QC, Yurtisigi MK, Sarker SK, Gupta A, Butler WH, Musfeldt JL (2013) Chemical tuning of the optical band gap in spinel ferrites: $CoFe_2O_4$ vs $NiFe_2O_4$. Appl Phys Lett 103:2–6. https://doi.org/10.1063/1.4818315
63. Sivula K, van de Krol R (2016) Semiconducting materials for photoelectrochemical energy conversion. Nat Rev Mater 1:15010. https://doi.org/10.1038/natrevmats.2015.10
64. Peeters D, Taffa DH, Kerrigan MM, Ney A, Jöns N, Rogalla D, Cwik S, Becker H-W, Grafen M, Ostendorf A, Winter CH, Chakraborty S, Wark M, Devi A (2017) Photoactive Zinc ferrites fabricated via conventional CVD approach. ACS Sustain Chem Eng 5:2917–2926. https://doi.org/10.1021/acssuschemeng.6b02233
65. Guijarro N, Bornoz P, Prévot M, Yu X, Zhu X, Johnson M, Jeanbourquin X, Le Formal F, Sivula K (2018) Evaluating spinel ferrites MFe_2O_4 ($M = Cu$, Mg, Zn) as photoanodes for solar water oxidation: prospects and limitations. Sustain Energy Fuels 2:103–117. https://doi.org/10.1039/C7SE00448F

Chapter 4
Chemical and Mechanical Properties of Ferrites

This chapter will present the most common aspects of the chemical and mechanical behaviors of spinel ferrites. The chemical versatility of the spinel structure is demonstrated, along with the doping and substoichiometry allowed by these materials. Surface functionalization and superficial charge are discussed. Aspects of the utilization of spinel ferrites as catalysts are considered, with a particular focus on Fenton and photo-Fenton reactions. Mechanical properties of ferrites are examined, and their sintering behavior is analyzed.

4.1 Introduction

The family of spinel ferrites draws most attention due to their magnetic and electrical properties, which were dealt with in the previous chapter. Nevertheless, other properties must be considered for the application of these materials in an industrial setting. The chemical properties of the final material are a crucial concern during the production of spinel ferrites, given the vast array of possibilities allowed by the flexibility of the spinel structure. Furthermore, the practical uses of ferrites are limited by their mechanical properties. The mechanical requirements of an application must match those allowed by the material. This chapter will address these two aspects of spinel ferrites.

4.2 Chemical Properties

The family of spinel ferrites shows great chemical versatility, which is one of the reasons for the variety of behaviors presented by these materials. The spinel structure offers excellent flexibility in its possible cation distribution. The divalent and trivalent cations are easily substituted by metals of equivalent charge, as shown in the previous

S. Arcaro and J. Venturini, *Modern Ferrites in Engineering*, Topics in Mining, Metallurgy and Materials Engineering, https://doi.org/10.1007/978-3-030-78988-6_4

chapters. Furthermore, the structure can also accommodate a large degree of oxygen vacancies. The properties and performance of the material are undoubtedly affected by the chemical composition of the final product.

One of the systems that have been intensely studied with respect to the effect of oxygen vacancies is zinc ferrite. Bulk $ZnFe_2O_4$ displays a normal spinel structure, with the Zn^{2+} ions in the tetrahedral site. As Zn^{2+} has no unpaired electrons, it does not exhibit an intrinsic magnetic moment. Therefore, bulk zinc ferrite displays a long-range antiferromagnetic below 10.5 K and a (mostly) paramagnetic character above this temperature, owing to the diamagnetic Zn^{2+} ion [1]. Nevertheless, nearly all nanosized zinc ferrites show ferromagnetism at room temperature. The observed magnetization is usually attributed to a partial inversion of the structure, giving rise to strong superexchange interactions between tetrahedral and octahedral Fe^{3+} ions.

This phenomenon has been widely observed for $ZnFe_2O_4$ prepared at reduced oxygen pressures. Philip and collaborators [2] have shown that the hysteresis curve of zinc ferrite undergoes a magnetic transition from para- to ferromagnetic when annealed in a vacuum at 1000 °C. Simultaneously, the authors show that the unit cell contracts at the same temperature, which was reported as a consequence of cationic rearrangement in the tetrahedral sites; larger Zn^{2+} moves toward the more spacious octahedral sites, while the trivalent ion occupies the smaller tetrahedral sites, resulting in the observed contraction. A more inverse character should translate into higher magnetization, as observed in their study.

Ayyappan et al. [3] studied a similar vacuum-annealed $ZnFe_2O_4$ system. While samples treated under air show a paramagnetic behavior, ferrites prepared under vacuum show very strong magnetization. Interestingly, when reannealed in air, the saturation of vacuum samples decreases from 62 to 1.5 emu.g^{-1}. Mössbauer and Raman data suggest that the oxygen vacancies created during the vacuum treatment cause a redistribution between the cationic interstitial sites.

However, Rodríguez Torres and collaborators reported in an in-depth study [4] that the inversion degree might not account for the magnetism observed in defect-rich $ZnFe_2O_4$. The authors have shown via X-ray magnetic circular dichroism (XMCD) experiments that octahedral Fe^{3+} is disproportionately responsible (x_B/x_A of 2.2) for the overall magnetism of zinc ferrites grown at low oxygen partial pressures. The large magnetic moment is thus assigned to the ferromagnetic coupling between octahedral iron atoms. The authors assign the breaking of the original antiferromagnetic ordering to three main mechanisms: A-site overpopulation, generation of octahedral vacancies, and the presence of oxygen vacancies between octahedral cations. The strong magnetization of these materials is, therefore, a manifestation of defect-induced magnetism.

The same structural flexibility that enables a wide range of substitution and stoichiometries also allows the facile doping of spinel ferrites. Doped analogs of these materials are widely reported in the literature, be it via the partial replacement of divalent [5, 6] or trivalent cations [7] or through aliovalent doping [8]. In the latter case, some of the original cations may change their oxidation state to satisfy the condition of crystalline charge neutrality. In a study on Mo^{6+}-doped cobalt ferrite,

Dwivedi et al. [9] have demonstrated that their material shows signatures of a ferroelectric behavior while also showing a magnetic character. The authors propose that the hexavalent cation enters the structure via the tetrahedral sites, converting the Fe^{3+} cations into Fe^{2+}. The observed ferroelectricity was assigned to the unfilled *d*-bands of molybdenum.

In order to avoid the creation of ferrous ions, Heiba and collaborators [10] prepared vanadium-doped cobalt ferrite with two stoichiometries: $CoFe_{2\text{-}x}V_xO_4$ and $CoFe_{2\text{-}1.67x}V_xO_4$. In the latter, each V^{5+} ion would replace 1.67 Fe^{3+} ions, thus maintaining charge neutrality with the creation of iron vacancies. Their studies show that vanadium occupies the octahedral site at low concentrations, with some migration toward the tetrahedral sublattice at higher ratios. Doping shows an effect on the coercivity of the material, mostly caused by the shift in cobalt positions. The iron-deficient sample shows a more substantial degradation in its magnetic properties, as would be expected.

Rare-earth (RE) doping is of particular interest in the technological development of spinel ferrites. Given that these ferrites are considered magnetically soft when compared to rare-earth magnets, the insertion of atoms with *f*-orbitals could have interesting effects on the coercivity of the final compounds. Indeed, gains in coercivity have been reported in the literature for RE-doped ferrites up to a critical doping fraction [11–13]. On the other hand, the magnetization of the doped analogs is virtually inescapably poorer than that of the pure version. The replacement of trivalent iron accounts for the loss in magnetization.

Mariosi and collaborators [14] studied a lanthanum-doped cobalt ferrite system prepared via a sol–gel method. The authors observed that the initial addition of lanthanum causes a contraction in the unit cell, which was assigned to the creation of defects when the large lanthanide ion enters the crystalline structure. Further increases in dopant fraction do not affect this parameter any further. Doping also led to gains in the surface area of the particles, from 74 to 110 $m^2.g^{-1}$, which was seen as a consequence of smaller crystallite sizes. Decreases in band gap, coercivity, and saturation were observed for the doped systems. The latter two were explained as a consequence of the smaller crystallites presented by the doped samples. Figure 4.1 shows the magnetic hysteresis curves of the studied materials and the relation between crystallite size and maximal magnetization.

Despite promising results, there is still great controversy in the literature as to what extent spinel ferrites can be factually doped by lanthanides. Tsvetkov et al. [15] have discussed in a recent work the subject of whether larger lanthanide ions can substitute Fe^{3+}. The authors prepared cobalt ferrites modified with different amounts of Ce(III), Nd(III), and Dy(III) to assess the effects of these ions in the structure of the products. The addition of rare-earths inexorably leads to decreases in crystallite size, as also observed in the previously discussed work. The authors assign this observation to the suppression effect of the larger ions on the crystallization of the spinel ferrite. This effect was more substantial at higher dopant contents. Neutron diffraction studies conducted for up to 50 h for each sample have shown no evidence of cell parameter variation upon doping, which was seen as a clear indication that the lanthanides were not effectively inserted into the spinel lattice. The inversion

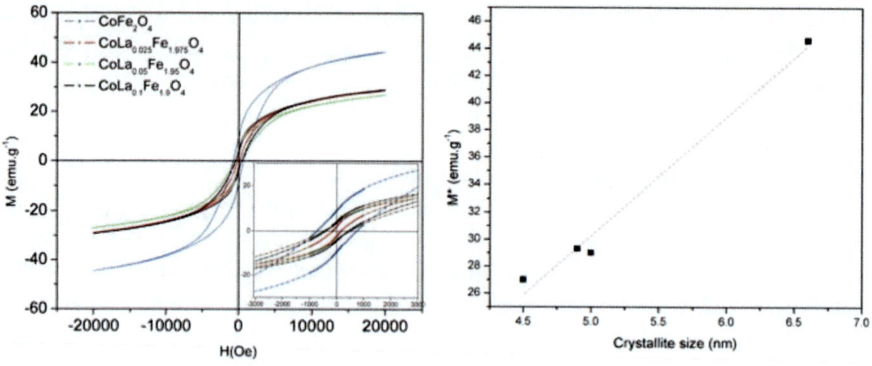

Fig. 4.1 Magnetic hysteresis behavior of lanthanum-doped cobalt ferrite nanoparticles. The relation between crystallite size and maximal magnetization is shown on the right (Reprinted from [14] with permission from Elsevier)

degree calculated via X-ray and neutron diffractograms also remained practically unaffected by the presence of the dopants. The researchers viewed these results as cabal evidence of the impossibility of lanthanide-doping. Instead, they proposed that the rare-earth ions are either distributed in crystalline phases of concentration below the detection limit of their techniques or in amorphous phases on the surface of the ferrite crystals. The effects on the magnetic and electrical properties would thus be caused by the presence of these secondary phases and not by actual doping.

Other than direct modifications of the crystalline structure, the properties of spinel ferrites may also be tuned via superficial modification. The surface of ferrites can be functionalized in a variety of ways. Amines, silanes, and inorganic acids are a few of the groups that have been attached to the surface of magnetic particles [16–18]. A few selected examples of moieties anchored to ferrites can be seen in Fig. 4.2.

Polymers are also commonly grafted onto ferrite particles, particularly in the field of magnetic adsorbents. An example of the preparation of polymer-modified spinel

Fig. 4.2 Selection of groups that may be successfully attached to spinel ferrites. (1) Trimethoxy silane, (2) diphenylphosphine, (3) dopamine 1, 2-diol, (4) phosphoric acid, (5) carboxylic acids, (6) amines, (7) alcohols, and (8) triphosphine (Reproduced from [19] with permission from Elsevier)

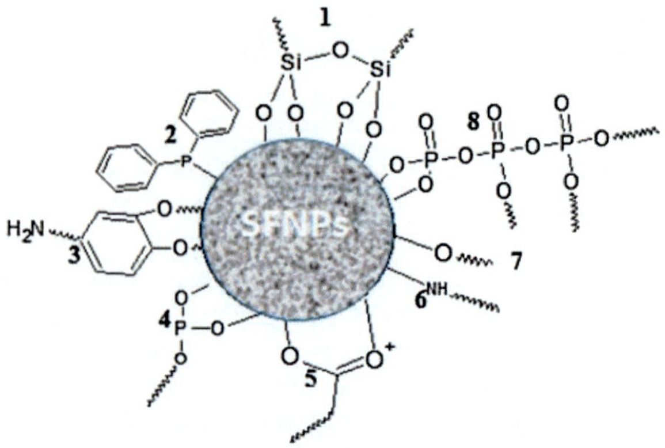

ferrites can be seen in Fig. 4.3, from the work of Mahdavian and Mirrahimi [20]. The authors took advantage of the hydroxyl groups present on the surface of magnetite to attach 3-aminopropyltriethoxysilane to the particles. The dangling amino group was further reacted with acryloylchloride to yield the amide. The vinyl moiety was reacted with acrylic acid, creating the polymeric layer attached to the magnetic core. The acidic hydrogen atoms in polyacrylate may undergo further reactions to yield the corresponding salts.

Besides the most apparent chemical consequences of surface functionalization, the effects of attaching functional groups may run much deeper. Vestal and Zhang have studied the impact of a wide range of surface-coordinated ligands on the magnetic properties of manganese ferrite [21]. The spinel nanoparticles were modified with a series of benzoic acid ligands (para-substituted with H, CH_3, Cl, NO_2, and OH) and benzene ligands substituted with COOH, SH, NH_2, OH, and SO_3H. Regardless

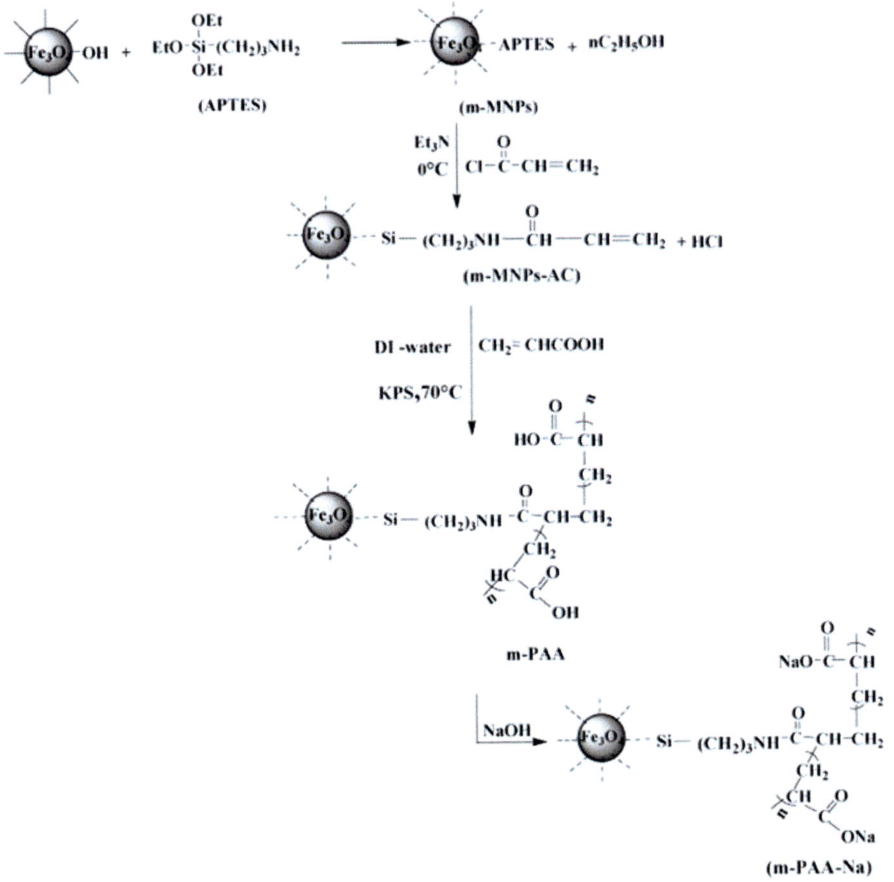

Fig. 4.3 Scheme of the chemical steps in the anchoring of polyacrylate onto magnetite particles (Reprinted from [20] with permission from Elsevier)

of the surface groups, coercivities decreased up to 48.3% while saturation magnetization values increased. A trend of larger decreases in coercivity for ligands with a stronger p-acceptor character was observed. The authors argue that larger crystal field stabilization energy (CFSE) splitting is generated by these ligands, leading to lower spin–orbit coupling and consequently lower coercivities. The most potent effects were observed for the smallest analyzed particles, as the surface contribution is much more pronounced in smaller crystallites.

Another important class of surface attachment to ferrite particles is molecules with biomedical significance. The magnetic cores are usually modified with a polysaccharide (e.g., chitosan, dextran) onto which the biomolecules are grafted [22]. Examples of these molecules can be seen in Fig. 4.4; these include enzymes, immunoglobulins, and nucleic acids, among others. The magnetic character of these modified ferrites has been used in improved detection methods, such as in novel protocols for the

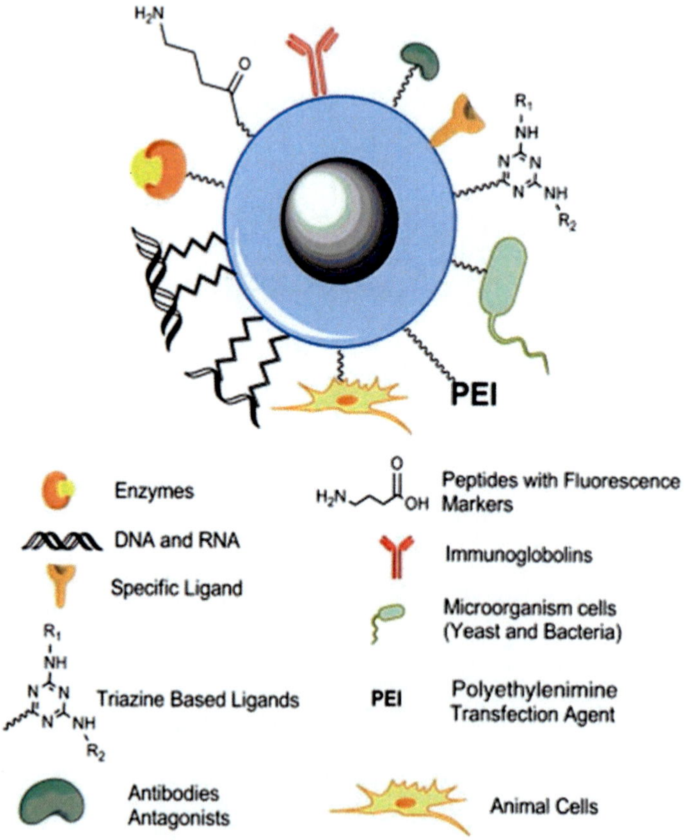

Fig. 4.4 Surface functionalization of spinel ferrites with a focus on groups of biomedical importance (Reprinted from [22] with permission from Elsevier)

detection of low-concentration analytes, from disease biomarkers [23] to viral RNA [24].

In the absence of functionalization, the surface of ferrites will behave in an aqueous solution as would any other particle: by abstracting H^+ or OH^- ions from the medium in order to reach chemical equilibrium. This process results in the creation of charged particles, which increases the colloidal stability of the system via coulombic repulsion. The generation of an electrical double layer around each charged component also ensures colloidal stability. The net charge in each particle depends mainly on the pH of the solution and the chemical characteristics of the particle.

At low pHs, the oxide particles become positively charged. Since each component has a positive charge, there is strong repulsion between particles. As the magnitude of this force is usually greater than that of the attractive van der Waals interactions, the system does not agglomerate. The same process takes place at higher pHs, albeit involving the opposite charges. In alkaline media, the particles become negatively charged via the incorporation of OH^- ions, but the same coulombic repulsion guarantees colloidal stability. Nevertheless, there is a pH value in which the negative and positive superficial charges cancel out, and the particles are effectively neutral. This condition is called point of zero charge (PZC or pH_{PZC}) or isoelectric point. At this pH, there is no repulsion to overcome the pull of van der Waals forces and the colloidal system will most likely agglomerate. Given its consequences, the point of zero charge is usually one to be avoided in most practical applications.

Colloidal stability is most commonly analyzed via the measurement of the zeta potential (z) of the system at different pHs. This parameter indicates the potential in the slipping plane separating the rigid layer from the diffuse layer relative to a point in the bulk liquid. As the zeta potential is caused by the charges in the particle, it is a widely used method for charge quantification. The isoelectric point is defined as the pH in which z equals zero.

Most common pristine ferrites exhibit an isoelectric point in the range between pH 5 and 8 [25–27]. Examples of zeta potential curves for $NiFe_2O_4$, $CoFe_2O_4$, and Fe_3O_4 are shown in Fig. 4.5, which was adapted from the work of Barale and collaborators [28]. Their work studied the conditions in which ferrites may deposit as corrosion products in pressurized water reactors (PWR), a type of nuclear reactor. The colloidal ferrites may become radioactive via neutron activation in the reactors, which might lead to radiation hotspots if they accumulate in certain parts of the circuit. The authors analyzed the effect of lithium and borate adsorption on the isoelectric point of ferrites in an attempt to avoid the agglomeration of undesired ferrites. The results show that the presence of borate reduces the isoelectric point of the three studied materials at 25 °C. An interesting thermal trend was also observed: While higher temperatures decreased the pH_{PZC} of $CoFe_2O_4$, nickel ferrite showed the opposite behavior, with higher isoelectric points.

The superficial state of ferrites is also crucial in adsorption processes. Adsorption is the adhesion of atoms or molecules to the surface of a material. This phenomenon occurs in an attempt to minimize the surface energy of the adsorbent. Atoms in the interface are not as stable as those in the bulk material; therefore, they are able to interact strongly with dissolved compounds (adsorbates) in order to minimize the free

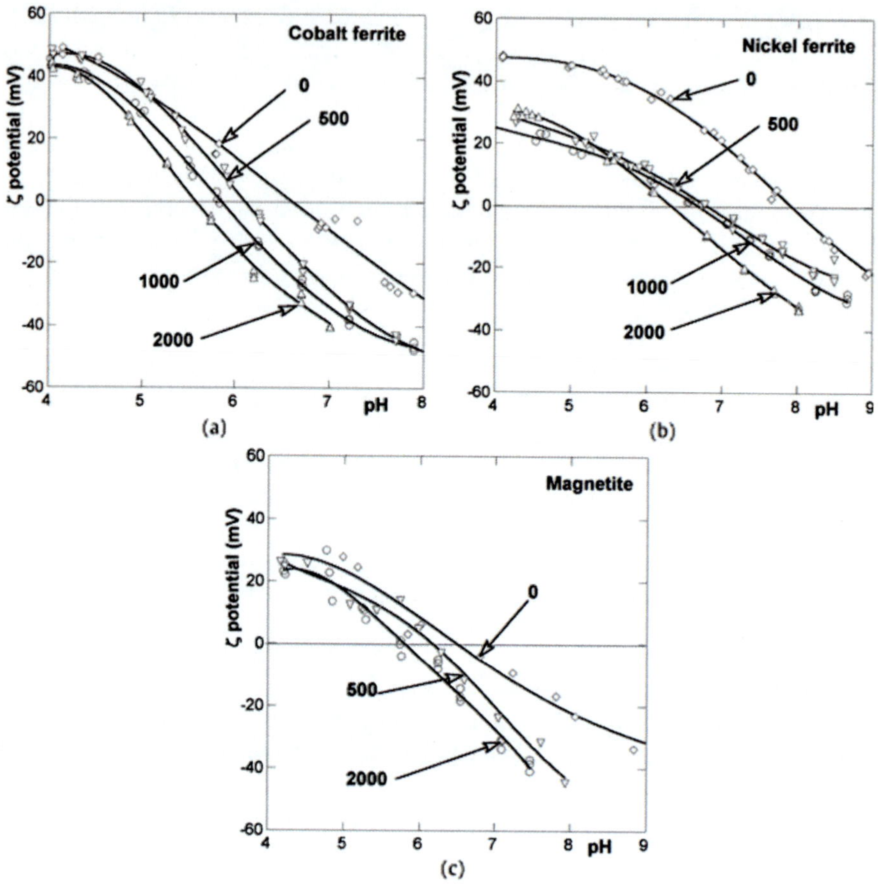

Fig. 4.5 Zeta potential (z) curves of cobalt ferrite, nickel ferrite, and magnetite in 10^{-4} M KCl solutions at 25 °C for different borate concentrations: 0, 500, 1000, and 2000 mg.L^{-1} (Reprinted from [28] with permission from Elsevier)

energy of the system. Adsorption is widely applied in separation processes in the industry. In this field, ferrites are particularly interesting due to the ease with which the adsorbent can be removed from the suspension; the application of a magnetic field is enough for the complete separation of these particles from the fluid.

A work by Konicki and coauthors [29] displays the capacity of ferrites in removing organic compounds from solution and the delicate interplay between pH and the efficiency of the process. $ZnFe_2O_4$ powders were prepared via a microwave-assisted hydrothermal method for posterior adsorption tests using Acid Red 88 as the adsorbate. The isoelectric point of the ferrite was determined via the pH drift method to be at approximately 8 (Fig. 4.6). Above this pH, the adsorption capacity of zinc ferrite was also shown to be minimal with regard to the organic dye. On the one hand, experiments carried out at the lower end of pH range show the largest capacities,

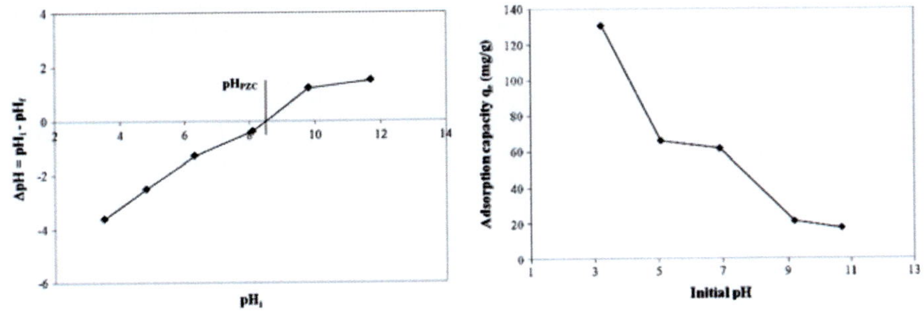

Fig. 4.6 Isoelectric point and adsorption capacity as a function of the pH for a $ZnFe_2O_4$/Acid Red 88 system (Reprinted from [29] with permission from Elsevier)

with over 100 $mg_{dye}/g_{ferrite}$, a mass efficiency of over 10%. The pH trend is explained by the charge state of both adsorbent and adsorbate. Acid Red 88 is an acid dye with pK_a below zero; it is always negatively charged in the studied range. On the other hand, zinc ferrite has a net positive charge in pH values below its pH_{PZC}. Therefore, the adsorbate and adsorbent have opposite charges in acidic pH, which ensures coulombic attraction and enhances the adsorption capacity of the system.

Besides its direct utilization in separation operations, adsorption is a crucial first step in catalytic processes that employ spinel ferrites. These have proven useful as heterogeneous catalysts in a myriad of reactions, such as dehydrogenation, oxidation, and alkylation, among others [30]. After the initial adsorption of the reactants onto the surface of ferrites, the transition metals in the spinel structure decrease the energy of the reaction intermediates, accelerating the reaction or increasing selectivity toward one of the possible products.

Spinel ferrites have performed well as catalysts in the oxidation of volatile organic compounds. For instance, Liu et al. [31] have studied the performance of manganese-substituted magnetite in the decomposition of formaldehyde, a compound whose emission is a relevant problem given its links to several types of cancer. XPS experiments show that the surface states of the manganese atoms are a mixture of +3 and + 4 valences. The study shows that the presence of manganese reduces the temperature of the events observed in the temperature-programmed reduction (TPR) curves of the catalysts. This facilitated reduction was linked to improved catalytic performance at lower temperatures in the oxidation of formaldehyde toward CO_2. Furthermore, the catalyst also showed good stability after a few cycles.

The presence of spinel ferrites may also aid reduction processes. $CuFe_2O_4$ may be successfully applied in the reduction of nitrophenols, a group of water pollutants with very complex treatment. Feng and collaborators [32] have shown that copper ferrite was able to convert nitrophenols to aminophenols, which are critical raw materials in the chemical industry. A 4-aminophenol conversion efficiency of 95% was observed after 40 s in the presence of $NaBH_4$. No similar effect was observed when Fe_2O_3 or Fe_3O_4 replaced $CuFe_2O_4$ in the system; therefore, the effect was assigned to the presence of copper. As adsorption of the pollutant was practically negligible after 15 min, the authors proposed that copper transfer electrons from BH_4^- to the nitrophenols, leading to the production of aminophenols.

The same compound has been utilized in water–gas shift (WGS) reactions, in which water and carbon monoxide produce hydrogen gas and CO_2. This reaction is also crucial in manufacturing several simple molecules (e.g., ammonia, methanol). Estrella and collaborators [33] conducted an in-situ analysis of the role played by $CuFe_2O_4$ as a catalyst in WGS systems. The results of H_2 and CO_2 production are shown in Fig. 4.7. XRD analysis indicates that WGS reactions take place above temperatures in which copper ions leave the structure, forming a Cu/Fe_3O_4 system. The authors propose that while metallic copper is the essential component, the oxide plays an active role. The barrier for water dissociation is much lower in the oxygen vacancies of oxides than in metallic copper. Therefore, this step likely takes place on the surface of the former.

Fenton reactions deserve particular attention in the field of catalysis due to their utilization in wastewater decontamination. Fenton's reagent—named after Henry John Horstman Fenton, who developed it at the end of the nineteenth century—is a solution of hydrogen peroxide and ferrous ion that generates free radicals. In this process, Fe^{2+} ions are oxidized to Fe^{3+} by the peroxide, generating a hydroxyl ion and a hydroxyl radical. In turn, the ferric ions are reduced by the peroxide, creating H^+ and a hydroperoxyl radical. The overall effect is the disproportionation of hydrogen peroxide, with the iron catalyst being continually regenerated. Due to the high rate of recombination of the free-radicals, a variation of the standard procedure was

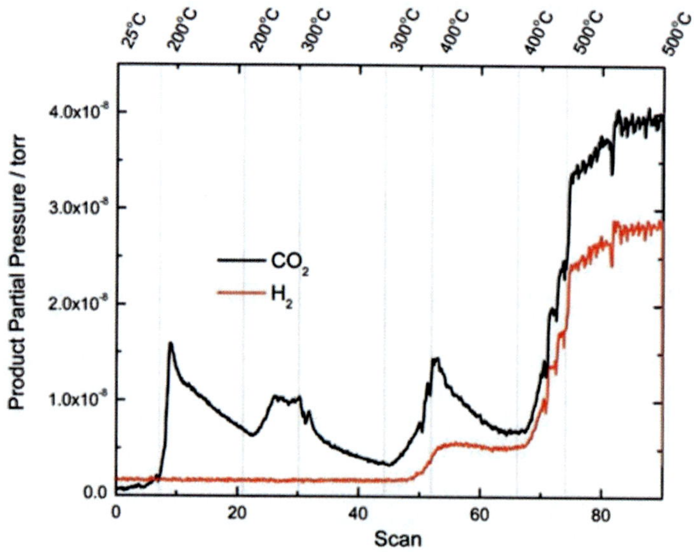

Fig. 4.7 Generation of CO_2 and H_2 after exposure of copper ferrite to a mixture of carbon monoxide and water (Reproduced with permission from [33]. Copyright [2009] American Chemical Society)

developed—the so-called photo-Fenton process, which uses UV light to accelerate the decomposition of the peroxide [34].

$$
\begin{array}{c}
Fe^{2+} + H_2O_2 \rightarrow Fe^{3+} + OH^- + HO^\bullet \\
Fe^{3+} + H_2O_2 \rightarrow Fe^{2+} + H^+ + HOO^\bullet \\
\hline
2H_2O_2 \rightarrow H_2O + HO^\bullet + HOO^\bullet
\end{array}
$$

The importance of these reactions comes from the oxidizing power of the products. Both hydroxyl and hydroperoxyl radicals are strong oxidants known to mineralize several organic compounds. For this reason, Fenton technology is widely applied as an advanced oxidation process for the polishing of wastewater contaminated with non-biodegradable organic compounds. In its usual setting, water treatment via homogeneous Fenton processes has major drawbacks. First, the pH of the solvent needs to be fixed at acidic pHs (2 to 3); otherwise, the concentration of Fe^{2+} and Fe^{3+} becomes unbalanced, with the consequent slowdown in decomposition rates [35]. Second, the generation of an iron sludge at the bottom of the reactor adds to the costs of the process, with the loss of catalyst. Furthermore, the utilization of homogeneous Fenton processes in the polishing of wastewater requires a secondary method to remove the iron atoms from the processed liquid.

The utilization of ferrites solves most of these problems. Ferrites naturally possess trivalent iron in their structure and, despite most of the members of this family not having Fe^{2+} in their structure, several other divalent metals may take its place in this catalytic process. An example of the proposed mechanisms for the generation of free radicals via heterogeneous photo-Fenton technology is shown in Fig. 4.8.

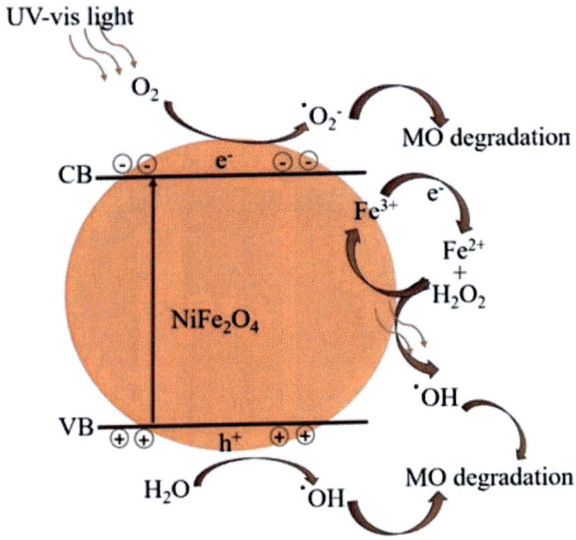

Fig. 4.8 Proposed mechanisms of the heterogeneous photo-Fenton degradation of methylorange (MO) on the surface of $NiFe_2O_4$ (Reproduced from [36] with permission from Elsevier)

Control over the adsorption of reagents onto the surface of ferrites may also speed up the reaction due to the higher local concentration of the pollutant of interest. Above all, the utilization of heterogeneous Fenton technology (in which the catalyst is not in the same phase as the reagents) simplifies the purification process. While a chemical process was necessary to remove iron species from the discharged water, the application of an electromagnetic field efficiently recovers the magnetic catalyst from the reactor. The recovered spinel ferrite may subsequently be reactivated and reutilized in further reactional cycles with little capacity loss.

Several spinel ferrites already show great promise for application in Fenton technology. For instance, Zhou et al. [37] studied the performance of an $MnFe_2O_4$/graphene composite as a heterogeneous catalyst in the decomposition of ammonia. A degradation of 92% was observed after 10 h with visible light irradiation of a 50.0 mg.L^{-1} NH_3 solution at pH 10.5 in the presence of 1.0 mmol.L^{-1} of peroxide. Instead of nitrates or nitrites, N_2 was identified as the degradation product. The pH was found to be very important in the determination of the degradation rates; these increase vastly at pHs above 9. The authors attribute this effect to the species present in the system. The pKa of NH_4^+ is 9.3; since only ammonia can be decomposed by hydroxyl radicals, the reaction only takes off under conditions in which the neutral species is prevalent.

Liu and collaborators [38] studied Fenton processes with the substitution of peroxide by oxalic acid, which may also generate OH• radicals. The authors used nickel ferrite as a heterogeneous photo-Fenton catalyst for the degradation of rhodamine B. The system reaches its maximum efficiency in the decomposition of the organic dye only in the presence of nickel ferrite, oxalic acid, and light, thus showing that the reaction proceeds via Fenton processes. The cycling behavior of the catalyst was also analyzed. The catalyst was magnetically removed from the system and subsequently introduced in another batch with the same initial concentration of reagents. The decolorization curves obtained in each of the catalytic cycles can be seen in Fig. 4.9. Nickel ferrite showed excellent stability after seven cycles, with a

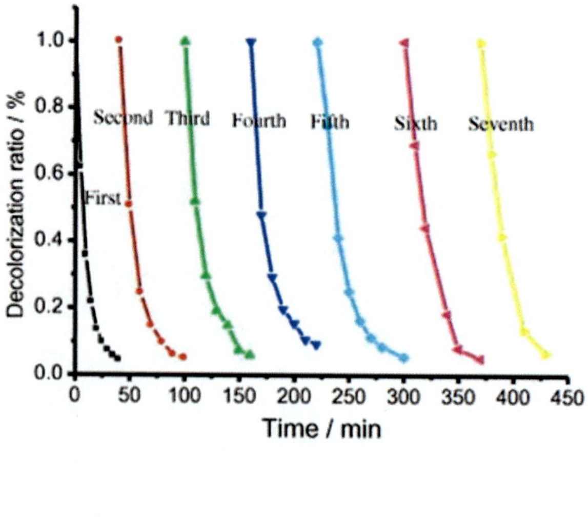

Fig. 4.9 Cycle stability of $NiFe_2O_4$ in the heterogeneous photo-Fenton decomposition of rhodamine B. $C_{oxalic\ acid} = 1.0$ mM, $W_{catalyst} = 0.20$ g, $C_{RhB} = 10.0$ mg/L, total volume = 50.0 mL (Reprinted from [38] with permission from Elsevier)

performance loss of less than 5% between the first and the last run.

4.3 Mechanical Properties

A vital consideration in choosing a material for any project is the mechanical requirements in the desired application. In the case of ferrites, for instance, their application in magnetomechanical sensors requires adequate mechanical strength for an appropriate utilization. High hardness values are also interesting for any magnetic material since they naturally attract other magnetic particles, possibly damaging the surface of the original material during impact. As usual for ceramic materials, ferrites tend to be both hard and brittle, which are important characteristics to be considered for their commercial application.

Spinel ferrites usually have theoretical densities of approximately 5.2 g.cm^{-3}, which is a relatively elevated value when compared to other typical engineering ceramics, such as Al_2O_3 or SiC. As full densification is hardly achieved, most ceramic bodies composed of ferrites reach lower densities. Commercial ferrite ceramics are usually reported to present between 4.7 and 4.8 g.cm^{-3}. The densities of fired spinel ferrites show a strong dependence on the temperature used during sintering. Gutiérrez-López et al. [39] analyzed the effect of this thermal treatment on the final density of NiZn ferrites prepared by powder injection molding and conventional uniaxial pressing. The comparison between the different methods and temperatures is shown in Fig. 4.10. The authors show that higher temperatures always lead to better densification in both processing methods. Furthermore, powder injection is also shown to yield higher densities when compared to uniaxial pressing. The researchers relate these results to the more homogeneous grain size and scarcer porosity present in the powder-injected samples.

Microwave sintering may be used in order to attain better densification of ferrites. This technique also usually requires lower temperatures, which is very interesting when considering the cost of industrial production. Tsay et al. [40] have shown these

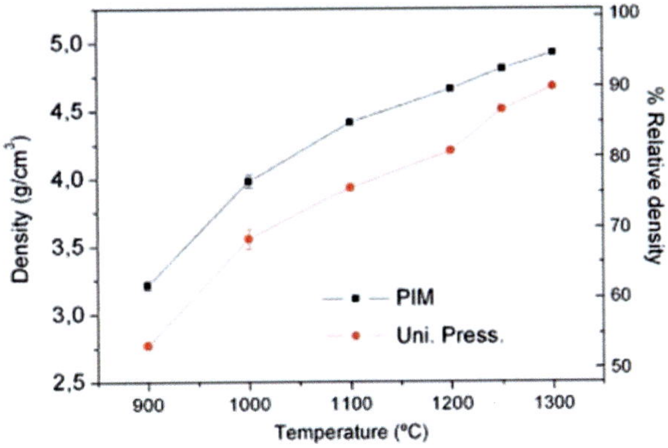

Fig. 4.10 Density and relative density of NiZn ferrites prepared via uniaxial pressing and powder injection molding sintered at different temperatures (Reprinted from [39] with permission from Elsevier)

advantages in the sintering of NiCuZn ferrite. The authors have achieved very high densification (<95%) after 10 min of microwave sintering at 800 °C, as can be seen in Fig. 4.11. Compared to conventional sintering, MW sintering can yield much denser materials at shorter times and lower temperatures. Several properties, such as the permeability of the sintered materials, also show improvement.

Spark plasma sintering (SPS) has also been used as a means to improve the densification of ferrite ceramics. As is the case of microwave-assisted sintering, this technique may yield denser ceramics at lower temperatures. Song et al. [41] have recently achieved almost full densification of NiZn ferrite via the application of SPS at a temperature of 925 °C. Similar results have also been reported for nanosized cobalt ferrite [42]. Spark plasma sintered pellets were produced from heat-treated $CoFe_2O_4$ precursors. The authors reported relative densities of 98% for ceramics sintered via SPS at 1000 °C. Similar sintering behavior is also shown in the literature for several other ferrites [43–45].

The mechanical hardness of spinel ferrites is also strongly influenced by the processing method. Nevertheless, these materials are usually regarded as relatively hard. Many ferrites commonly take Vickers hardness values between 5

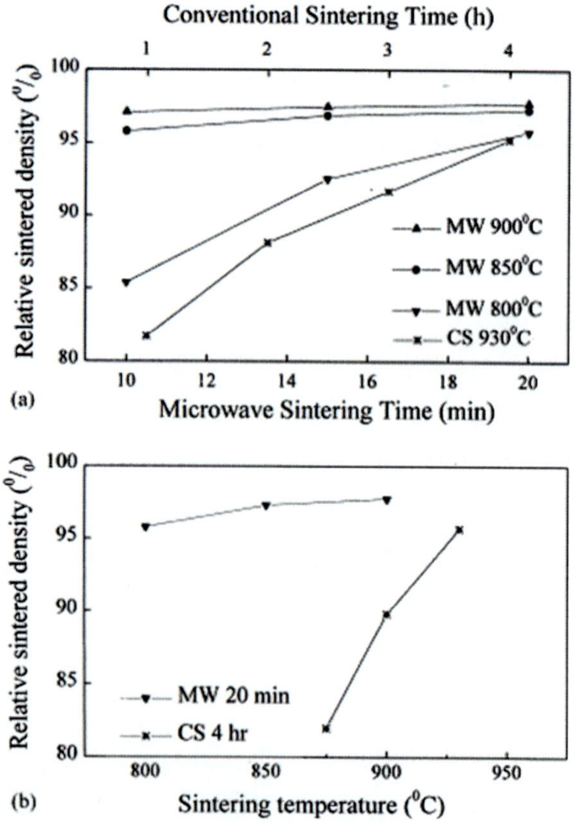

Fig. 4.11 Relative density of NiCuZn ferrites sintered via microwave and conventional sintering (Reprinted from [40] with permission from Elsevier)

and 8 GPa. Rezlescu and collaborators [46] analyzed the effect of rare-earth doping on the electrical and mechanical behavior of a nickel-zinc ferrite, namely $Ni_{0.7}Zn_{0.3}Fe_{1.98}R_{0.02}O_4$, where R denotes a rare-earth metal. The authors have found that the addition of small amounts of dopants may lead to improvements in the mechanical properties due to the crystallization of secondary phases.

Venturini et al. [47] have studied the impact of different solvents in the sol–gel synthesis of cobalt ferrite on the mechanical hardness of sintered samples. The authors report that powders prepared in a 1:1 mixture of water and isopropanol exhibit higher Vicker hardness than their counterparts prepared in either pure water or an 80% isopropanol mixture. This improved mechanical performance was assigned to the morphology of the grains, which were much smaller and more angular when prepared in the intermediary solvent. Firing temperature also displayed significant influence, with samples treated at 1200 °C showing higher hardness than those fired at 1150 °C, regardless of the preparation method.

A related study also further shows the effect of temperature on the mechanical properties of $CoFe_2O_4$ [48]. Cobalt ferrite samples were initially treated at 750, 800, 850, and 900 °C, pressed into pellets, and fired at 1150 and 1200 °C. The results show that the Vickers hardness of the samples again is maximized with increasing firing temperatures. A Hall–Petch plot of the samples (Fig. 4.12) shows a tendency of increasing hardness with decreasing particle sizes, albeit not the expected linear relationship. The authors assign this deviation from linearity to the heterogeneity in particle size observed in the samples.

Dias and Moreira [49] have also assessed the effect of temperature on the mechanical hardness of ferrites, most specifically hydrothermal nickel-zinc ferrite sintered at 1100, 1200, 1300, and 1400 °C. For similar grain sizes, higher firing temperatures yielded improved hardness. However, this enhancement is quickly lost at longer treatment times. Grain growth is reported as the cause for this phenomenon, as the

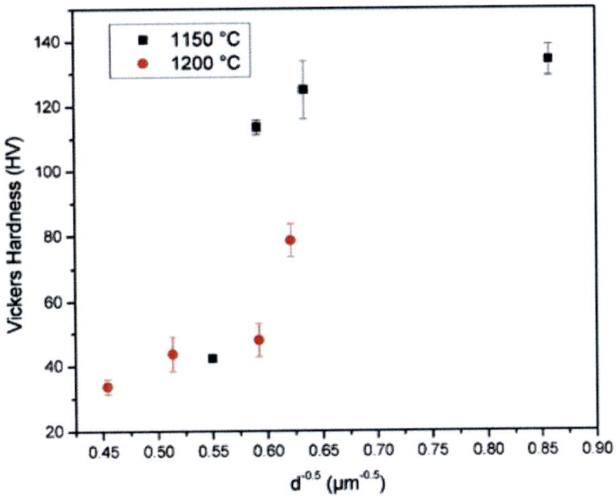

Fig. 4.12 Hall–Petch plot of $CoFe_2O_4$ sintered samples fired at 1150 °C and 1200 °C (Reprinted from [48] with permission from Elsevier)

samples behave in accordance with the Hall–Petch relation. For samples with larger grain sizes, hardness decreases systematically.

References

1. Schiessl W, Potzel W, Karzel H, Steiner M, Kalvius GM, Martin A, Krause MK, Halevy I, Gal J, Schäfer W, Will G, Hillberg M, Wäppling R (1996) Magnetic properties of the $ZnFe_2O_4$ spinel. Phys Rev B 53:9143–9152. https://doi.org/10.1103/PhysRevB.53.9143
2. Philip J, Gnanaprakash G, Panneerselvam G, Antony MP, Jayakumar T, Raj B (2007) Effect of thermal annealing under vacuum on the crystal structure, size, and magnetic properties of $ZnFe_2O_4$ nanoparticles. J Appl Phys 102:54305. https://doi.org/10.1063/1.2777168
3. Ayyappan S, Raja SP, Venkateswaran C, Philip J, Raj B (2010) Room temperature ferromagnetism in vacuum annealed $ZnFe_2O_4$ nanoparticles. Appl Phys Lett 96:143106. https://doi.org/10.1063/1.3374332
4. Rodríguez Torres CE, Pasquevich GA, Zélis PM, Golmar F, Heluani SP, Nayak SK, Adeagbo WA, Hergert W, Hoffmann M, Ernst A, Esquinazi P, Stewart SJ (2014) Oxygen-vacancy-induced local ferromagnetism as a driving mechanism in enhancing the magnetic response of ferrites. Phys Rev B 89:104411. https://doi.org/10.1103/PhysRevB.89.104411
5. Kardile HJ, Somvanshi SB, Chavan AR, Pandit AA, Jadhav KM (2020) Effect of Cd2+ doping on structural, morphological, optical, magnetic and wettability properties of nickel ferrite thin films. Optik (Stuttg) 207:164462. https://doi.org/10.1016/j.ijleo.2020.164462
6. Kavitha S, Kurian M (2019) Effect of zirconium doping in the microstructure, magnetic and dielectric properties of cobalt ferrite nanoparticles. J Alloys Compd 799:147–159. https://doi.org/10.1016/j.jallcom.2019.05.183
7. Naik MM, Naik HSB, Nagaraju G, Vinuth M, Vinu K, Rashmi SK (2018) Effect of aluminum doping on structural, optical, photocatalytic and antibacterial activity on nickel ferrite nanoparticles by sol–gel auto-combustion method. J Mater Sci Mater Electron 29:20395–20414. https://doi.org/10.1007/s10854-018-0174-y
8. Farea AMM, Kumar S, Batoo KM, Yousef A, Lee CG (2009) Alimuddin, influence of the doping of Ti4+ ions on electrical and magnetic properties of Mn1+xFe2−2xTixO4 ferrite. J Alloys Compd 469:451–457. https://doi.org/10.1016/j.jallcom.2008.01.139
9. Dwivedi GD, Tseng KF, Chan CL, Shahi P, Lourembam J, Chatterjee B, Ghosh AK, Yang HD, Chatterjee S (2010) Signature of ferroelectricity in magnetically ordered Mo-doped $CoFe_2O_4$. Phys Rev B 82. https://doi.org/10.1103/PhysRevB.82.134428
10. Heiba ZK, Mohamed MB, Ahmed SI (2017). Cation distribution correlated with magnetic properties of cobalt ferrite nanoparticles defective by vanadium doping. J Magn Magn Mater 441:409–416. https://doi.org/10.1016/j.jmmm.2017.06.021
11. Prathapani S, Vinitha M, Jayaraman TV, Das D (2014) Effect of Er doping on the structural and magnetic properties of cobalt-ferrite. J Appl Phys 115:17A502. https://doi.org/10.1063/1.4854915
12. Lakshmiprasanna HR, Jagadeesha Angadi V, Rajesh Babu B, Pasha M, Manjunatha K, Matteppanavar S (2019) Effect of Pr3+-doping on the structural, elastic and magnetic properties of Mn–Zn ferrite nanoparticles prepared by solution combustion synthesis method. Chem Data Collect 24:100273. https://doi.org/10.1016/j.cdc.2019.100273
13. Kahn ML, Zhang ZJ (2001) Synthesis and magnetic properties of $CoFe_2O_4$ spinel ferrite nanoparticles doped with lanthanide ions. Appl Phys Lett 3651:4–7. https://doi.org/10.1063/1.1377621
14. Mariosi FR, Venturini J, da Cas Viegas A, Bergmann CP (2020) Lanthanum-doped spinel cobalt ferrite ($CoFe_2O_4$) nanoparticles for environmental applications. Ceram Int 46:2772–2779. https://doi.org/10.1016/j.ceramint.2019.09.266

15. Tsvetkov M, Milanova M, Ivanova I, Neov D, Cherkezova-zheleva Z (2019) Phase composition and crystal structure determination of cobalt ferrite, modified with Ce, Nd and Dy ions by X-ray and neutron diffraction. J Mol Struct 1179: 233–241. https://doi.org/10.1016/j.molstruc.2018.07.083
16. Bohara RA, Thorat ND, Yadav HM, Pawar SH (2014) One-step synthesis of uniform and biocompatible amine functionalized cobalt ferrite nanoparticles: a potential carrier for biomedical applications. New J Chem 38:2979–2986. https://doi.org/10.1039/C4NJ00344F
17. Liu P, Yao Z, Zhou J (2015) Controllable synthesis and enhanced microwave absorption properties of silane-modified $Ni_{0.4}Zn_{0.4}Co_{0.2}Fe_2O_4$ nanocomposites covered with reduced graphene oxide. RSC Adv 5:93739–93748. https://doi.org/10.1039/C5RA18668D
18. Xu P, Zeng GM, Huang DL, Feng CL, Hu S, Zhao MH, Lai C, Wei Z, Huang C, Xie GX, Liu ZF (2012) Use of iron oxide nanomaterials in wastewater treatment: A review. Sci Total Environ 424: 1–10. https://doi.org/10.1016/j.scitotenv.2012.02.023
19. Kefeni KK, Mamba BB, Msagati TAM (2017) Application of spinel ferrite nanoparticles in water and wastewater treatment: a review. Sep Purif Technol 188:399–422. https://doi.org/10.1016/J.SEPPUR.2017.07.015
20. Mahdavian AR, Mirrahimi MA-S (2010) Efficient separation of heavy metal cations by anchoring polyacrylic acid on superparamagnetic magnetite nanoparticles through surface modification. Chem Eng J 159: 264–271. https://doi.org/10.1016/j.cej.2010.02.041
21. Vestal CR, Zhang ZJ (2003) Effects of surface coordination chemistry on the magnetic properties of $MnFe_2O_4$ spinel ferrite nanoparticles. J Am Chem Soc 125:9828–9833. https://doi.org/10.1021/ja035474n
22. Dias AMGC, Hussain A, Marcos AS, Roque ACA (2011) A biotechnological perspective on the application of iron oxide magnetic colloids modified with polysaccharides. Biotechnol Adv 29:142–155. https://doi.org/10.1016/j.biotechadv.2010.10.003
23. He Y, Wang Y, Yang X, Xie S, Yuan R, Chai Y (2016) Metal organic frameworks combining $CoFe_2O_4$ magnetic nanoparticles as highly efficient SERS sensing platform for ultrasensitive detection of N-terminal pro-brain natriuretic peptide. ACS Appl Mater Interfaces 8:7683–7690. https://doi.org/10.1021/acsami.6b01112
24. Somvanshi SB, Kharat PB, Saraf TS, Somwanshi SB, Shejul SB, Jadhav KM (2020) Multifunctional nano-magnetic particles assisted viral RNA-extraction protocol for potential detection of COVID-19. Mater Res Innov, 1–6. https://doi.org/10.1080/14328917.2020.1769350
25. de Vicente J, Delgado AV, Plaza RC, Durán JDG, González-Caballero F (2000) Stability of cobalt ferrite colloidal particles. Eff of pH and App Magn Fields, Langmuir. 16:7954–7961. https://doi.org/10.1021/la0003490
26. Niu Z, Zhang S, Zhu L (2018) A study of biochemical route on construction of waste battery ferrite applying for nickel removal. Environ Sci Pollut Res 25:21577–21588. https://doi.org/10.1007/s11356-018-2057-4
27. Jadhav SA, Somvanshi SB, Khedkar MV, Patade SR, Jadhav KM (2020) Magneto-structural and photocatalytic behavior of mixed Ni–Zn nano-spinel ferrites: visible light-enabled active photodegradation of rhodamine B. J Mater Sci Mater Electron 31:11352–11365. https://doi.org/10.1007/s10854-020-03684-1
28. Barale M, Lefèvre G, Carrette F, Catalette H, Fédoroff M, Cote G (2008) Effect of the adsorption of lithium and borate species on the zeta potential of particles of cobalt ferrite, nickel ferrite, and magnetite. J Colloid Interface Sci 328: 34–40. https://doi.org/10.1016/j.jcis.2008.09.007
29. Konicki W, Sibera D, Mijowska E, Lendzion-Bieluń Z, Narkiewicz U (2013) Equilibrium and kinetic studies on acid dye Acid Red 88 adsorption by magnetic $ZnFe_2O_4$ spinel ferrite nanoparticles. J Colloid Interface Sci 398:152–160. https://doi.org/10.1016/j.jcis.2013.02.021
30. Kharisov BI, Dias HVR, Kharissova OV (2019) Mini-review: Ferrite nanoparticles in the catalysis. Arab J Chem 12:1234–1246. https://doi.org/10.1016/j.arabjc.2014.10.049
31. Liu P, He H, Wei G, Liang X, Qi F, Tan F, Tan W, Zhu J, Zhu R (2016) Effect of Mn substitution on the promoted formaldehyde oxidation over spinel ferrite: catalyst characterization, performance and reaction mechanism. Appl Catal B Environ 182:476–484. https://doi.org/10.1016/j.apcatb.2015.09.055

32. Feng J, Su L, Ma Y, Ren C, Guo Q, Chen X (2013) $CuFe_2O_4$ magnetic nanoparticles: a simple and efficient catalyst for the reduction of nitrophenol. Chem Eng J 221:16–24. https://doi.org/10.1016/j.cej.2013.02.009
33. Estrella M, Barrio L, Zhou G, Wang X, Wang Q, Wen W, Hanson JC, Frenkel AI, Rodriguez JA (2009) In situ characterization of $CuFe_2O_4$ and Cu/Fe_3O_4 Water–Gas Shift Catalysts. J Phys Chem C 113:14411–14417. https://doi.org/10.1021/jp903818q
34. Venturini J (2019) Application of ferrite nanoparticles in wastewater treatment BT. In: Kopp Alves A (ed) Nanomaterials for eco-friendly applications. Springer, Cham, pp 141–151. https://doi.org/10.1007/978-3-030-26810-7_9
35. Wang Y, Zhao H, Li M, Fan J, Zhao G (2014) Magnetic ordered mesoporous copper ferrite as a heterogeneous Fenton catalyst for the degradation of imidacloprid. Appl Catal B Environ 147:534–545. https://doi.org/10.1016/j.apcatb.2013.09.017
36. Ahmed S, Ahmad Z (2020) Development of hexagonal nanoscale nickel ferrite for the removal of organic pollutant via Photo-Fenton type catalytic oxidation process. Environ Nanotechnology, Monit Manag 14:100321. https://doi.org/10.1016/j.enmm.2020.100321
37. Zhou Y, Xiao B, Liu S-Q, Meng Z, Chen Z-G, Zou C-Y, Liu C-B, Chen F, Zhou X (2016) Photo-Fenton degradation of ammonia via a manganese–iron double-active component catalyst of graphene–manganese ferrite under visible light. Chem Eng J 283:266–275. https://doi.org/10.1016/j.cej.2015.07.049
38. Liu S-Q, Feng L-R, Xu N, Chen Z-G, Wang X-M (2012) Magnetic nickel ferrite as a heterogeneous photo-Fenton catalyst for the degradation of rhodamine B in the presence of oxalic acid. Chem Eng J 203: 432–439. https://doi.org/10.1016/j.cej.2012.07.071
39. Gutiérrez-López, J, Rodriguez-Senín E, Pastor JY, Paris MA, Martín A, Levenfeld B, Várez A (2011) Microstructure, magnetic and mechanical properties of Ni–Zn ferrites prepared by powder injection moulding. Powder Technol 210:29–35. https://doi.org/10.1016/j.powtec.2011.02.008
40. Tsay CY, Liu KS, Lin TF, Lin IN (200) Microwave sintering of NiCuZn ferrites and multilayer chip inductors. J Magn. Magn Mater 209:189–192. https://doi.org/10.1016/S0304-8853(99)00684-8
41. Song S, Song Q, Li J, Mudinepalli VR, Zhang Z (2014) Characterization of submicrometer-sized NiZn ferrite prepared by spark plasma sintering. Ceram Int 40:6473–6479. https://doi.org/10.1016/j.ceramint.2013.11.099
42. Millot N, Le Gallet S, Aymes D, Bernard F, Grin Y (2007) Spark plasma sintering of cobalt ferrite nanopowders prepared by coprecipitation and hydrothermal synthesis. J Eur Ceram Soc 27:921–926. https://doi.org/10.1016/j.jeurceramsoc.2006.04.141
43. Penchal Reddy M, Shakoor RA, Mohamed AMA, Gupta M, Huang Q (2016) Effect of sintering temperature on the structural and magnetic properties of $MgFe_2O_4$ ceramics prepared by spark plasma sintering. Ceram Int 42:4221–4227. https://doi.org/10.1016/j.ceramint.2015.11.097
44. Jenuš P, Topole M, McGuiness P, Granados-Miralles C, Stingaciu M, Christensen M, Kobe S, Žužek Rožman K (2016) Ferrite-based exchange-coupled hard–soft magnets fabricated by spark plasma sintering. J Am Ceram Soc 99:1927–1934. https://doi.org/10.1111/jace.14193
45. Mesbahinia A, Almasi-Kashi M, Ghasemi A, Ramazani A (2020) FORC investigation of Co-Ni bulk ferrite consolidated by spark plasma sintering technique. J Magn Magn Mater 497:165976. https://doi.org/10.1016/j.jmmm.2019.165976
46. Rezlescu E, Rezlescu N, Popa PD, Rezlescu L, Pasnicu C (1997) The influence of R_2O_3 (R = Yb, Er, Dy, Tb, Gd, Sm and Ce) on the electric and mechanical properties of a nickel-zinc ferrite. Phys Status Solidi 162:673–678. https://doi.org/10.1002/1521-396X(199708)162:2%3c673::AID-PSSA673%3e3.0.CO;2-A
47. Venturini J, Wermuth TB, Machado MC, Arcaro S, Alves AK, da Cas Viegas A, Bergmann CP (2019) The influence of solvent composition in the sol-gel synthesis of cobalt ferrite ($CoFe_2O_4$): a route to tuning its magnetic and mechanical properties. J Eur Ceram Soc 39:3442–3449. https://doi.org/10.1016/j.jeurceramsoc.2019.01.030
48. Caldeira LE, Guaglianoni WC, Venturini J, Arcaro S, Bergmann CP, Bragança SR (2020) Sintering-dependent mechanical and magnetic properties of spinel cobalt ferrite ($CoFe_2O_4$)

ceramics prepared via sol-gel synthesis. Ceram Int 46:2465–2472. https://doi.org/10.1016/j.ceramint.2019.09.240

49. Dias A, Moreira RL (1999) Chemical, mechanical and dielectric properties after sintering of hydrothermal nickel–zinc ferrites. Mater Lett 39:69–76. https://doi.org/10.1016/S0167-577X(98)00219-5

Chapter 5
Synthesis

Different technologies are being evaluated to produce spinel ferrites. Knowledge about existing synthesis methods allows greater control over the structure, and consequently, the desired properties. This chapter will describe the mechanisms of the main synthesis methods of spinel ferrites and analyze the approach in selected articles.

5.1 Introduction

Diversified synthesis methods were developed to produce ferrites with reasonable control of shape, composition, size and size distribution. In general, the synthesis methods can be divided into two categories: "bottom-up" and "top-down." In "bottom-up" methods, the ions are chemically combined to form the particles. In "top-down" synthesis techniques, the materials are diminished to form tiny particles.

There are appreciable synthesis techniques for the "bottom-up" option: sol–gel, co-precipitation, thermal decomposition, hydrothermal or solvothermal, sonochemical, vapor deposition, microwave-assisted, microemulsion, and others. Only the mechanical milling techniques and pulsed laser ablation are common "top-down" synthesis methods.

Commonly, the "bottom-up" methods are preferred due to a good shape and size distribution control. There is no universal method of synthesis of ferrites. Each synthesis method has its advantages and disadvantages.

5.2 Standard Ceramic Techniques

Standard ceramic techniques are mostly used in the industrial production of ferrites. The method is an example of a "top-down" approach and it is relatively simple and cost-effective. The method involves heating a mixture of oxides and carbonates or

S. Arcaro and J. Venturini, *Modern Ferrites in Engineering*, Topics in Mining, Metallurgy and Materials Engineering, https://doi.org/10.1007/978-3-030-78988-6_5

hydroxides to obtain the desired crystalline phase (solid-state reaction) to synthesize the spinel.

In this process, initially, the raw materials are milled to produce a tiny material. Subsequently, the powders are heat treated. Usually, high temperatures and long thermal treatment times are used to obtain the structure of interest. However, this method results in a coarse-grained product. It is crucial to make sure that the starting materials are intimately mixed at the atomic or ionic level before reaction, so crystallization can occur at a lower temperature, resulting in smaller grains. This effect can be achieved by high-energy milling. The method enables the production of nanosized powders after long milling times.

The synthesis parameters are fundamental and control the final properties of the obtained materials. $Ni_{0.5}Zn_{0.5}Nd_xFe_{2-x}O_4$ ($0 \leq x \leq 0.1$) ferrites [1] were synthesized by the standard ceramic method. To reduce the raw materials' particle size and achieve good homogeneity, the powders were mixed and ground for 3 hours. A homogeneous mixture of ceramic oxide powders was sintered in three steps. First, the pre-sintering of fine powder was heat-treated at 900°C for 16 hours. In the second step, the sintered powder was again ground to reduce the particle size and again pre-sintered at 1100°C for 16 hours. After sintering at 1100°C, the powder was compacted into a pellet. The final sintering of pellets was conducted at 1250°C for 16 hours. The results show that the standard ceramic techniques were able to produce doped spinel ferrites. Nd-doped ferrites exhibit cubic spinel phase for $0.00 \leq x \leq 0.08$ with a single-phase present in the product. For $x = 0.1$, the splitting of the (311) reflection is observed along with main cubic spinel phase reflections. The variation in Mössbauer parameters indicates that the Nd-doped ferrites exhibit a chemical formula $(Zn_{0.5}Fe_{0.5-x}Nd_x)_A[Ni_{0.5}Fe_{1.5}]_BO_4$. In summary, $(Zn_{0.5}Fe_{0.5-x}Nd_x)_A[Ni_{0.5}Fe_{1.5}]_BO_4$ reflects the fact that the Nd atoms prefer the tetrahedral coordination, reducing the spinel inversion parameter [1].

Impurity-free $NiFe_2O_4$ has been synthesized using a planetary ball milling-assisted solid state reaction process [2]. A standard ceramic technique was used to obtain $NiFe_2O_4$ with a wide size distribution (35–85 nm) and spherical morphologies for smaller particles and polygons for larger ones. The authors observe that the increase in heat treatment temperature from 750 to 800 °C resulted in increased Ms from 33.4 to 42.7 emu g^{-1} and the average size change from 40.3–54.8 nm, and a decrease in corresponding coercivity (Hc) from 143.2 to 116.6 Oe.

While milling occurs, several chemical reactions take place, leading to a metastable product that needs a heat treatment to be stabilized. Figure 5.1 shows the schematic illustration of the formation process of $NiFe_2O_4$ nanoparticles.

On the other hand, in another study, the same milled precursors were calcined at 800°C for the same time; the corresponding determined values of Ms and Hc of the obtained $NiFe_2O_4$ particles were 46.5 emu g^{-1} and 107.6 Oe, respectively [3].

These results show the impact of the synthesis method on the Ms and Hc of $NiFe_2O_4$. The mechanical milling differences were sufficient to modify the structure of the materials, which consequently modified the final properties. Therefore, improving the synthesis technique is one of the required conditions in order to get better crystallinities and superior properties.

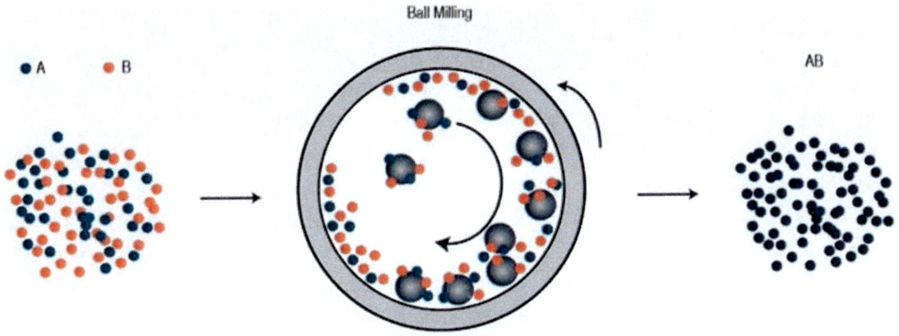

Fig. 5.1 Mechanical milling synthesis

In some cases, mechanical activation can be used to directly obtain the desired crystalline phase, i.e., without the need for thermal energy. The significant advantage of mechanical milling is that it offers the possibility of scaling up to tonnage quantities of materials for various applications, while its most serious drawback is the contamination of particles during milling. Figure 5.1 illustrates the mechanical milling synthesis.

The fundamental mechanism of size reduction in mechanical activation is the energy applied to the sample during impacts with the milling media. The particles are trapped between two colliding balls and then compressed into smaller pieces, as shown in Fig. 5.2. Initially, the particles slide past one another with minimum deformation and fracture, producing refined, irregularly shaped particles. Then, in the second step, the particles undergo elastic and plastic deformation. Finally, the particle undergoes fracture, which results in further fragmentation of the particles. As the fragments decrease in size, the tendency to form aggregates increases, and fracture resistance increases. Particle fineness approaches a limit as milling continues and maximum energy is expended [4].

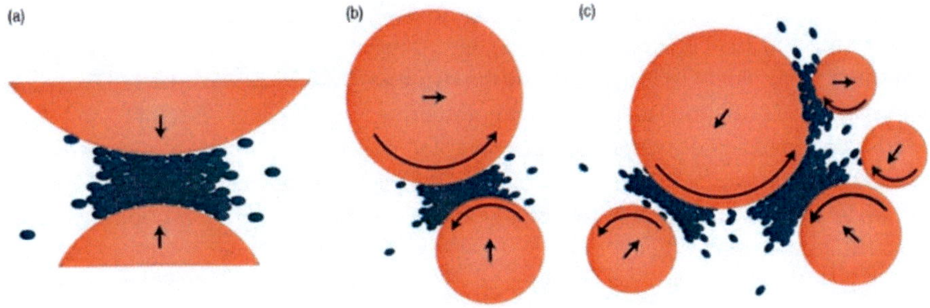

Fig. 5.2 Schematic diagram showing the different forms of impact during high-energy ball milling (**a**) Head-on impact (**b**) Oblique impact (**c**) Multi-ball impact

Several parameters can affect the powder's final constitution, among them, the type of mill, milling container, milling speed, milling time, grinding medium, process control agent (surfactant added to the powder during milling to inhibit agglomeration), and temperature of milling. These parameters can be tuned depending on the final properties expected of the powders.

By mechanosynthesis, large amounts of nanocrystalline ferrites can be obtained in a relatively short time. The nanocrystalline ferrites can be prepared by mechanosynthesis with a mean crystallites size of a few nanometers and a particle size of micrometers [5–8]. In the case of doped ferrites, one of the essential aspects is seeing how the dopant elements influence the powder structural and magnetic characteristics and their evolution when subjected to mechanical milling. Marinca and collaborators [6] studied a series of manganese-substituted nickel spinel ferrites, $Ni_{1-x}Mn_xFe_2O_4$ (x = 0, 0.3, 0.5, 0.7). These ferrites have been successfully synthesized both in polycrystalline and nanocrystalline/nanosized states by mechanical milling of ceramics obtained from stoichiometric mixtures of oxides. It was observed that cation substitution of Mn^{2+} for Ni^{2+} leads to an increase of magnetization and to the expansion of the crystal lattice. Mechanical milling is favorable to the synthesis of the manganese-substituted nickel ferrites in the nanocrystalline state, although, for milling times longer than a few minutes, a decrease of both magnetization and ordering temperature can be noticed. It was evidenced that in the ferrites with a large amount of Mn^{2+} cations a more rapid reduction of the crystallite size occurs. A reduction of the Néel temperature by more than 200°C is also observed upon increasing the Mn^{2+} content from x = 0 to 0.7. The magnetization of the mechanically milled samples tends to become difficult to saturate, and this behavior is assigned to the spin canting effect occurring at the grain surface. It was observed that the substitution of Mn^{2+} by Ni^{2+} limits the magnetic disorder.

Lysenko et al. [9] studied the microstructure of a mechanically activated Fe_2O_3-Li_2CO_3-ZnO mixture to produce lithium-zinc ferrites. High-energy planetary ball milling with steel grinding balls and a rotation speed of 2220 rpm for up to 60 minutes was used. It was shown that the composition of the mixture changes during the ball milling, which consists of decreasing α-Fe_2O_3 concentration and increasing Fe_3O_4 spinel phase. The spinel phase concentration increases with increasing milling time, while the Li_2CO_3 and ZnO concentrations remain unchanged. It was found that the milling leads to a decrease in the average particle size of the reagents and simultaneous formation of large-size agglomerates with denser structures and well-interlinked particles. It was established that the observed changes in microstructure and phase composition lead to an increase in the reactivity of the Fe_2O_3-Li_2CO_3-ZnO system and the acceleration of the chemical reaction between reagents obtained from the milling process.

5.3 Laser Ablation

Laser ablation is a deposition technique that uses a laser beam as an energy source for the evaporation of a material to be deposited. Because of the almost unlimited number of materials and various solvents suitable for laser ablation, it has been used to synthesize an enormous variety of advanced materials. Besides, picosecond lasers that operate at high power with a high repetition rate can significantly improve the yield of synthesized particles [10]. In this method, small-size particles are produced by irradiation of a solid surface with a liquid medium laser beam. In this synthesis technique, three critical stages are recognized; in the initial stage, the target material absorbs laser energy; after that, energy relaxation leads to the formation of heat waves that propagate inside the target material and the liquid. This results in material ejection and plume formation. In the second stage, the laser plume expands in the presence of a confining liquid environment. Finally, the plume mixes with the liquid and nanoparticle aggregation occurs. Short laser pulses favor the formation of tiny particles. The technique allows us to synthesize various particles, with size controlled by a laser-induced size reduction or by varying the liquid type. Besides, since such a synthesis technique is free of reducing agents, it is possible to synthesize particles with high purity. The deposited material's chemical and physical characteristics will depend on the substrate temperature, substrate crystal composition and structure, pressure, laser pulse repetition rate, laser beam targeting, species, gas pressure, and distance between the target and the substrate.

Various types of pulsed lasers can be used for the ablation of materials. Currently, the most used lasers are those of Nd:YAG and those of exciters. The Nd:YAG is a solid-state laser with a fundamental frequency of 1064 nm, but one can use folders, triplicates, and quadruplicates of the optical frequency to obtain wavelengths of 532, 355, and 266 nm, respectively. These lasers are widely used because they are relatively inexpensive, require little maintenance, do not present any divergence from the beam, and are rather small. Therefore, they can be easily incorporated in an ablation system. Figure 5.3 shows the Fe_3O_4 particles obtained by laser ablation.

The power threshold needed to create the plasma plume depends on the wavelength and pulse duration of the laser beam and the target species. ICCD imaging and optical emission spectroscopy analysis can be used to obtain information on the dynamics of the plasma plume and of its constituents together with the spatial distribution of the excitation temperature of atoms. The authors of this work [12] obtained cobalt ferrite with preferential crystallographic structure, controlled composition, and perpendicular magnetic anisotropy using laser ablation. Their properties are influenced by the experimental conditions and thus by the characteristics of the deposition plasma plume.

This method is used mainly due to the absence of ligands and precursor residues, producing nanoparticles with bare surfaces. This fact is very important in terms of applications in catalysis. It enables increased interaction between the catalyst and adsorbate.

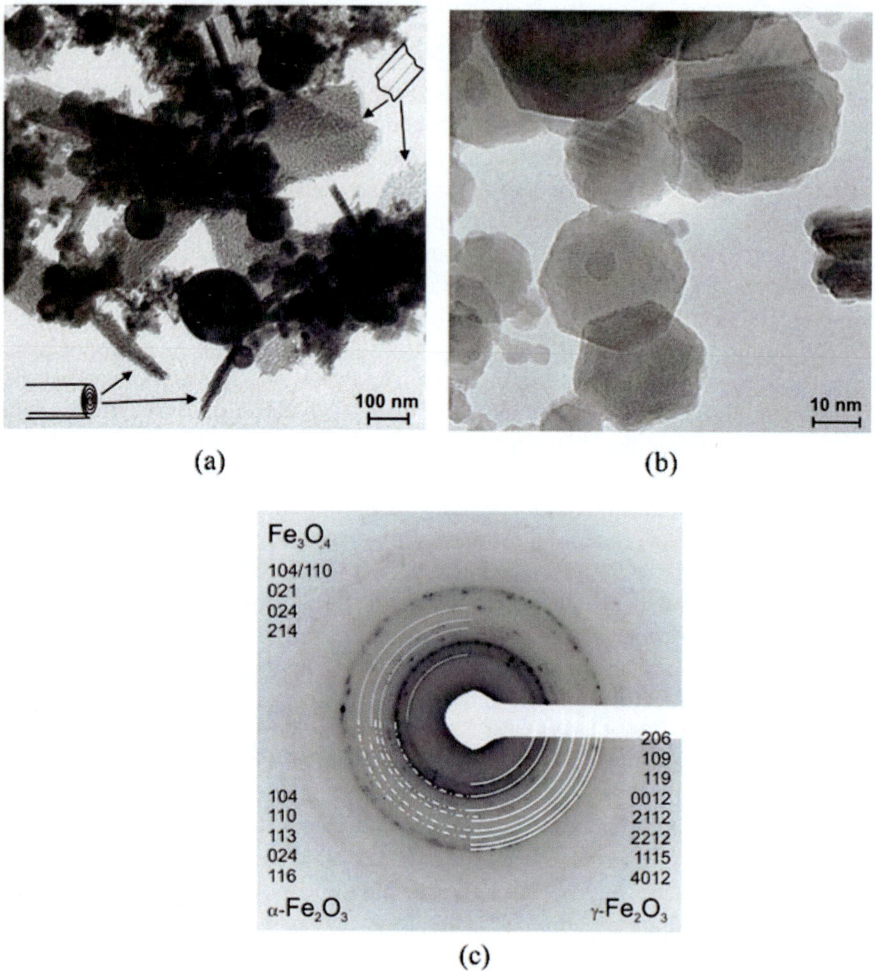

Fig. 5.3 Fe_3O_4 nanoparticles were obtained by laser ablation. Reprinted from [11] with permission from Elsevier

Small variations in the composition exposed to laser can tune different material properties. Titanium ferrites were synthesized by Shukla and collaborators [13], using liquid-assisted pulsed laser ablation followed by post-irradiation treatment in double-distilled water. With increases of titanium fraction in the target material, the optical band gap increases, while particle size, ferromagnetism, degree of spherical particle formation, and rate of agglomeration/aggregation are decreased. These trends indicate that one must choose a higher Ti/Fe ratio in order to get smaller-sized but less ferromagnetic nanoparticles, and vice versa. Higher ferromagnetism with larger-sized particles demonstrates the dominance of bulk phenomena over surface defects for the ferromagnetism of spins. Moreover, variation in laser pulse energy at a given

Ti/Fe ratio does not show much effect on the optical band gap, size, and magnetic properties of produced nanoparticles.

5.4 Solution Combustion Synthesis

A method for the preparation of magnetic particles which offers outstanding characteristics is the solution combustion synthesis. This technique was introduced for the preparation of oxide materials two decades ago [14]. Oxidizing reagents (nitrates, sulfates, carbonates, etc.) and an organic fuel or complexing agent (e.g., citric acid, urea, glycine, carbohydrazide, maleic hydrazine, sucrose) that act as a reducing reagent are solubilized in water. Different oxidizer-to-fuel ratios can be used to increase the temperature of the combustion reaction [15–17]. The solution is heated to the boiling point resulting in a rapid, self-sustaining reaction. The energy released during the exothermic reaction between the oxidizing reagents and the fuel during combustion can rapidly and intensely heat up the system, keeping it sufficiently heated for a prolonged period, even in the absence of an external thermal source. High temperatures are reached during the synthesis, which guarantees the crystallization and formation of the ceramic materials in a short time. Figure 5.4 shows an illustration of the solution combustion synthesis.

Solution combustion synthesis can produce fine and homogeneous particles by atomization of the liquid before reaction and by the explosive nature of the reaction itself. The large volume of gases generated in the reaction (CO_2, NO_x, H_2O) helps to cool the product quickly, thus preventing the growth of crystallites [19]. Besides, the gas release leads to formation of powders with a porous structure and good surface area. The pyrolyzed powder that is left may be the single-phase product but

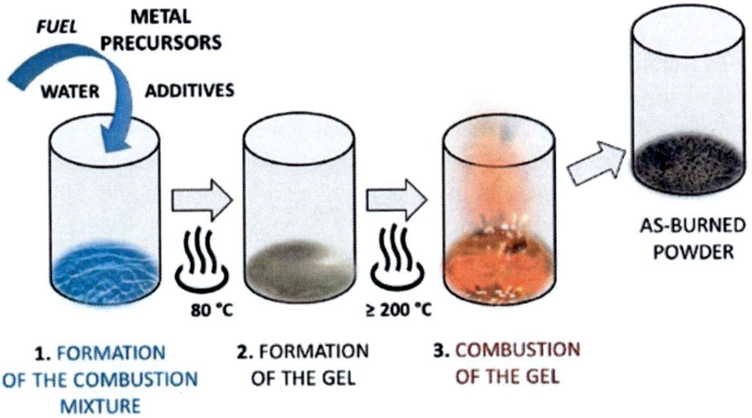

Fig. 5.4 Illustration of the solution combustion synthesis. Reprinted from [18] with permission from Elsevier and The Lancet

usually will be a combination of metal oxides as the ultimate result of the process. Homogeneity is achieved primarily because the system in solution is mixed on an atomic scale, and diffusion is therefore limited to the size of the liquid droplet before drying or combustion.

The main process parameters that directly interfere with the properties of the final products are the type of fuel, the fuel/oxidizer ratio, and the geometry of the reactor. On the other hand, the disadvantages of this process include strongly agglomerated or aggregated powders that may also exhibit carbon traces of the combustion reaction.

This synthesis method was utilized in the synthesis of magnesium ferrites ($MgFe_2O_4$) using iron nitrate and magnesium nitrate as precursors and glycine as fuel (30% fuel-deficient), without the necessity of subsequent heat treatments after synthesis [20, 21]. The authors obtained crystallite sizes between 18 and 61 nm with the most significant value of the specific surface area of 28 m^2/g. In the fuel-deficient sample, traces of a secondary phase, such as hematite, present in the other $MgFe_2O_4$ compositions were not detected. The authors observed that magnetic hysteresis curves indicated a ferromagnetic behavior of the samples at room temperature. The increase of Ms might have resulted from the increase of the particle size and improvement of crystallinity since coercivity does not vary significantly between the samples. The high-performance magnetic material containing $MgFe_2O_4$ spinel should prove useful in some interesting applications.

In the solution combustion synthesis of spinels, the enthalpy and the flame temperature generated during combustion principally govern the particle characteristics, such as crystallite size, surface area, nature of agglomeration, and the degree of spinel inversion [22]. These properties are predominantly tuned by controlling the chemical composition of the fuel and the ratio of fuel to oxidizer. The stoichiometry of the reactions is determined based on calculations for fuel-oxidizer mixtures in the chemistry of propellants, which considers the amount of oxidizing and reducing elements in the reaction. However, the stoichiometric ratio is not always the most efficient [22–24]. Thus, Zampiva and collaborators [23] evaluated the production of homogeneous $MgFe_2O_4$ nanoparticles with high surface area and different degree of spinel inversion. For this purpose, they studied the application of glycine and polyethylene glycol—200 molecular weight (PEG 200)—as fuels, as well as the influence of the fuel/oxidizer ratio in the solution combustion synthesis of $MgFe_2O_4$ catalyst nanoparticles.

The highest degree of spinel inversion is shown by the samples produced with PEG. In addition, the formation of a second phase, MgO, is observed. On the other hand, the use of glycine resulted in porous structures formed of high-purity nanoparticles. The fuel composition and the concentration during the solution combustion synthesis proved crucial for obtaining high-quality powders. Glycine presented elevated efficiency in the production of $MgFe_2O_4$ nanoparticles functionalized with PEG. The glycine-lean conditions provided the most excellent results, which guaranteed fast and complete combustion and resulted in high surface area and reduced particle size structures.

5.5 Sol–Gel

The sol–gel method is a chemical route used to synthesize colloidal suspensions of inorganic or hybrid materials. It is one of the most common methods for synthesizing complex metal oxides, metastable materials, and organic–inorganic hybrid compounds. The main advantages of this process are the low temperatures involved and high homogeneity. Although there are several approaches to the sol–gel method, there are basic steps common to all: A "sol" (colloidal solution) of the precursors gradually evolves into a gel by condensing the molecules in the mixture. The precursors are typically alkoxides or organic and inorganic salts of metals, dissolved in aqueous media or organic solvents. Catalysts, additives—such as surfactants and polymers—and gelling agents (e.g., carboxylic acids) may be added to the blend to direct the reaction toward desired final shapes and sizes.

To remove solvents and side products of this structure, a drying step is required, usually by heating the gel. This step strongly influences the final structure of the material; aerogels, for example, have their structure defined in the drying step, which occurs by the passage of supercritical fluids. For most materials, drying alone is not enough for a high-quality product. Further heat treatment is also required for complete polycondensation, raising the crystallinity of the final material.

Some of the advantages of the sol–gel synthesis method include the fact that it operates at low temperature and low cost, and does not need particular types of equipment. The reaction temperature in sol–gel usually varies between 25 and 200°C, and it is possible to synthesize particles with narrow size distribution and controlled shape. These advantages, coupled with its simplicity in terms of the synthesis procedures, make the sol–gel method very attractive. In addition, it is one of the preferred synthesis methods to control the composition, microstructure, purity, and shape of particles by adjusting different parameters such as sol concentration, stirring rate, and annealing temperature. Over the last few years, the sol–gel techniques have been used to prepare a variety of ferrites. It has been demonstrated that the sol–gel process offers considerable advantages such as better mixing of the starting materials and excellent chemical homogeneity in the final product. Moreover, the molecular level mixing and the tendency of partially hydrolyzed species to form extended networks facilitate the structure evolution, thereby lowering the crystallization temperature.

One of the sol–gel methods used to produce mixed oxides worth mentioning is the citrate–nitrate method, mainly due to its versatility. The process involves using inorganic nitrates as sources of cations and citric acid ($C_6H_8O_7$) as a mineralizer.

The presence of the organic compound fulfills a variety of functions. As a chelant, citric acid complexes the cations, increasing their solubility and, consequently, ensuring that precipitation of undesired oxides and hydroxides does not occur. Citric acid also helps in the better distribution of the cations within the formed reticule, guaranteeing the homogeneity of the formed material. This factor is of great importance when it comes to the synthesis of oxides of two or more metals. In addition, citric acid acts as a mineralizer, helping to form the incipient polymer network. The presence of the organic matrix ensures that, when nucleation occurs, the sites are numerous and well-distributed, guaranteeing small particle size.

Moreover, its third role in the synthesis is that of fuel. Under heating in the presence of nitrates, which are strong oxidizing agents, a rapid self-sustaining combustion reaction occurs in the xerogel. In this reaction, nitrates rapidly oxidize the organic matter contained in citric acid. In several cases, the heating caused by this autoignition is sufficient for the crystallization of the desired material. The release of gases during combustion can also lead to the generation of porous or spongy products, as desired for various applications. In addition, the speed with which this exothermic reaction occurs can generate thermodynamically metastable forms since kinetic conditions take over the process control. This is the case of several processes of production of cobalt ferrites, which do not have enough time to relax their structure due to the violence of the combustion.

The importance of controlling sol–gel parameters translates into the direct tuning of the spinel product's desired characteristics. Spinel cobalt ferrites ($CoFe_2O_4$) with fluctuating levels of substoichiometric were prepared by Venturini and his research group via sol–gel synthesis with different combinations of citric acid, dextrose, and PVP (polyvinylpyrrolidone) [25]. The results showed the possibility of tuning the magnetic and structural properties of the spinel product via control of the composition of the reaction medium. The synthesized materials showed up to 50% deficiency of the metal, which leaves the spinel lattice and forms Fe_2O_3 instead. The fuel/oxidant ratio and the thermal behavior of the xerogels exert a precise influence on the compositional variation of the prepared spinels.

As in combustion synthesis, it is also possible to use fuels in the sol–gel synthesis. These fuels play the role of chelating agents and can accelerate combustion and assist in the homogeneous distribution of cations in the reaction medium, favoring the final product. Glycine [26], urea [19, 27], maltose [28], and sucrose [29, 30] are examples of these fuels.

Besides the control of the parameters of the synthesis, the citrate–nitrate post-synthesis parameters are also very important and very versatile. For example, the heat treatment temperature should be thoroughly evaluated as it may directly influence the particle size of the formed ferrites [31–33]. Cobalt ferrite nanoparticles with an average particle size in the order of 10 nm were successfully synthesized via the sol–gel method under mild conditions. A marked tendency toward a normal spinel structure, unlike what is expected for bulk cobalt ferrite, was observed. Impedances studies demonstrated that the material displayed a metallic behavior at temperatures up to 100°C, above which the material became semiconducting, opening possibilities for novel advanced applications.

Figure 5.5 shows that the produced cobalt ferrite nanoparticles revealed a minute size of approximately 10 nm. The images indicate a high degree of aggregation, most likely due to the magnetic interaction between each grain. The particles tend toward a spherical shape and the average size of the particles is very homogeneous. The interplanar spacing correspondent to the (311), (220), and (111) crystalline planes of $CoFe_2O_4$ can be clearly identified, indicating that despite the mild conditions under

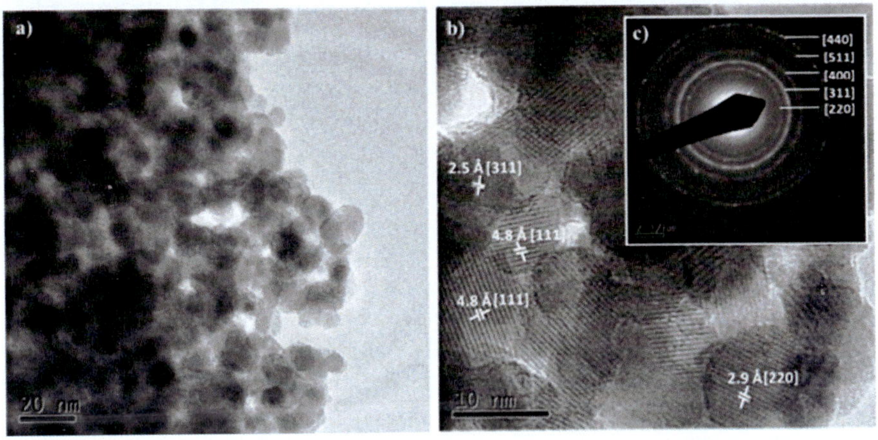

Fig. 5.5 nanoparticles produced revealed a minute size for the cobalt ferrite nanoparticles. Reproduced from [31] with permission from the The Royal Society of Chemistry

which the precursor gel was treated, $CoFe_2O_4$ nanoparticles with an elevated degree of crystallinity were produced.

Moreover, tuning the magnetic and mechanical properties of sintered ceramics can be performed with strict control of synthesis process parameters. Research [30, 34] shows that the reactional media (water, isopropanol, or mixtures) displays a significant influence on the morphology of the grains and on the final properties of the $CoFe_2O_4$.

Along with modifying the synthesis processes, researchers have also explored different doping possibilities in ferrites. Significant variation in structural and magnetic properties of these ferrites have been observed with doping. Furthermore, while there are several efforts toward synthesizing doped spinel ferrites, the main difficulty during the synthesis is the impurity phase (orthoferrite, hematite, metal monoxide) segregation [35–38]. However, due to the ease of parameter adjustment via sol–gel citrate–nitrate synthesis, these problems are meager. Cobalt ferrite was doped with lanthanum [32] and there was a significant variation in the size of the crystallite with an increase in the addition of the dopant. The insertion Cof lanthanum causes a complete rearrangement of the cationic sublattices of spinel ferrite, with a tendency for an inverse character when the dopant is added to the structure. In fact, the optical gap of the samples was reduced from 1.35 to 1.1 eV in the doped samples. In addition, the magnetic parameters of the samples were also strongly affected by the insertion of the lanthanide ion.

Single-phase, nanocrystalline nickel-zinc ferrite samples with nominal erbium doping were prepared nitrate citrate method. The incorporation of erbium into the system inhibits the grain growth and introduces lattice strain into the system. The particle size is drastically reduced from 22 to 8 nm in the samples. The room-temperature magnetic measurements reveal that the saturation magnetization (Ms) of Er^{3+}-substituted Ni-Zn ferrites decreases with the addition of the Er^{3+} ion within the lattice. This observation has been correlated with the rearrangement of cations due to Er^{3+} doping. The doping of ferrites provided an almost 63% enhancement in coercivity [38].

5.6 Electrospinning

Electrospinning is a versatile technique that uses electrostatic forces to produce fine fibers from a polymer solution. This method uses high voltage sources (5–50 kV) and low currents (0.5–1 μA) to produce fibers of reduced diameter when a jet of a fluid material is accelerated and drawn in the presence of an external electric field.

A typical electrospinning process basically consists of a high voltage source electrically connected to a fluid (polymer solution), a collector (where the fibers are deposited), and an infusion device (pump). Typically, the polymers are dissolved in solvents to form the polymer solution, which is introduced in the syringe. The solution is forced to flow via a needle, which is under the action of an external electric field, to a collector that can be flat and stationary or cylindrical and rotating. Figure 5.6 shows the electrospinning schematic system.

The application of a high electric voltage creates an electric field between the needle of the syringe and the collector. Under the influence of the electric field, a drop of the polymer solution at the tip of the needle is deformed in the shape of a cone. When the applied potential reaches a value necessary to overcome the surface tension of the polymer solution, a solution jet is ejected from the tip of the cone and accelerated by the electric field toward the collector. The jet pathway can be controlled by the electric field, which is responsible for lengthening the jet. As the jet travels in the air, the solvent evaporates making the jet thinner and giving rise to the fibers.

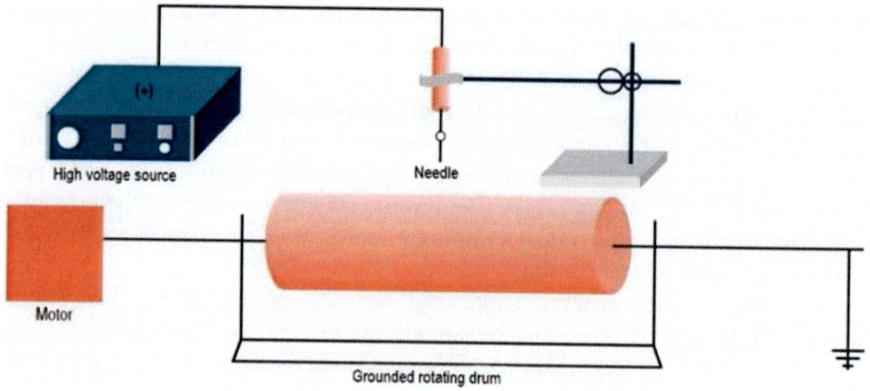

Fig. 5.6 Schematic of an electrospinning system

Fig. 5.7 Images of $La_{0.6}Sr_{0.4}Co_{0.8}Fe_{0.2}O_3$ fiber; (**a**) SEM image of electrospun fibers; (**b**) SEM image of fibers thermally treated at 1000°C; and (**c**) TEM image of fibers thermally treated at 1000°C. Reprinted (adapted) from [39] with permission from the American Chemical Society

The morphology and diameter of the fibers obtained by electrospinning are dependent on the intrinsic properties of the solution (polymer type, viscosity, elasticity, electrical conductivity of the solution, and polarity and surface tension of the solvent), operating conditions, electric field strength, distance between the needle and the collector, movement of the collector, feed rate of the polymer solution, and environmental conditions (temperature, humidity, and air velocity).

By appropriate manipulation of these parameters, one can obtain fibers of desired morphology and diameter. Lubini and collaborators [39, 40] studied fibers of $La_{0.6}Sr_{0.4}Co_{1-y}Fe_yO_3$ ($y = 0.2 - 1.0$) synthesized by electrospinning to produce promising materials for solid oxide fuel cell cathodes that operate at intermediate temperatures. Simple solvents (ethanol) and nitrate salts were used as raw materials for the fiber manufacturing process. The influence of Co and Fe contents in the structural and electrical properties of the prepared fibers was studied. The obtained fibers presented homogeneous distribution and average diameter of around 2 μm. Figure 5.7 shows that the obtained electrospun fibers are smooth and uniform, displaying a mean diameter in the range of 1.7 –2.6 μm. After thermal treatment at 1000°C (Fig. 5.7b), the mean diameter of the fibers decreased to ~1 μm for all compositions. This reduction occurs probably due to polymer decomposition and the usual shrinkage of ceramic materials after a thermal treatment process, producing a denser material and shorter fibers. The TEM image in Fig. 5.7c shows that thermally treated fibers are composed of interconnected agglomerated nanoparticles.

Among all the materials prepared by those authors, the $La_{0.6}Sr_{0.4}Co_{0.8}Fe_{0.2}O_3$ fibers obtained by electrospinning showed the most promising electrical and electrochemical properties for oxygen reduction reactions. Therefore, this material could be used in solid oxide fuel cell cathodes at intermediary temperatures.

5.7 Co-Precipitation

Co-precipitation is a synthesis method used to prepare multicomponent oxides via precursor formation, a pure or a variously complex multicomponent intermediate precipitate. In this synthesis method, the process depends on the ions in the starting materials. These ions are typically inorganic salts, precipitated by the addition of stoichiometric amounts of bases or acids. Moreover, the reaction medium usually is water. The precursor is irreversibly converted into oxide by thermal treatment. Co-precipitation allows an intimate mixing of all the desired ions in the oxide, maintaining chemical homogeneity on the molecular scale.

Co-precipitation reactions involve the simultaneous occurrence of nucleation, growth, coarsening, and agglomeration processes. Chemical reactions that result in low solubility products are generally chosen so that the solution quickly reaches a supersaturated condition.

The processes of nucleation and growth govern the particle size and morphology of products in precipitation reactions. Factors such as reaction rate, transport rate of reactants, removal, and redistribution of matter compete with thermodynamic influences in particle growth. The reaction and transport rates are affected by the concentration of reactants, temperature, pH, and order of introduction and degree of mixing of reagents. Reaction rates and impurities may influence the structure and crystallinity of particles. Factors such as supersaturation, nucleation and growth rates, colloidal stability, recrystallization, and aging processes also affect particle size and morphology. At low supersaturation, the particles are small, compact, and well-formed, and the shape depends on crystal structure and surface energies. At high supersaturation, large and dendritic particles form. At even higher supersaturation, smaller but compacted and agglomerated particles form, which means supersaturation shows predominant influence on the precipitate's morphology.

The main advantages of co-precipitation in relation to the other methods of synthesis are the evenly mixed reactants; raw materials without organic solvent; simple process; greater homogeneity and reactivity; high purity; the possibility of synthesis of new materials; nanometric particle size; narrow size distribution; control of particle size and distribution using adjustments in the synthesis.

Synthesis of ferrites by co-precipitation is among the most frequently applied techniques used to synthesize uniform nanoparticles. In this method, aqueous solutions, which contain the mixture of divalent and trivalent transition metals, are mixed. The synthesis method requires careful adjustment and control of pH in order to prepare high-quality materials. The pH of the solution is usually adjusted using ammonium hydroxide or sodium hydroxide solutions. Thereafter, the solution's vigorous stirring will follow in the absence or presence of heat and an inert atmosphere. Figure 5.8 shows the schematic illustration of the co-precipitation synthesis process.

The crystallinity of ferrites prepared by co-precipitation in aqueous solution is relatively low. Therefore, subsequent heat treatment is required to improve the degree of crystallinity of the particles. Another important aspect of the co-precipitation method is the high possibility of scaling up of the process. Furthermore, one of the desirable features of the co-precipitation method is that it is more environmentally friendly since only water is used for salt dissolution. The final effluent can be easily discharged after treatment for anion removal. In the co-precipitation method, control of size and size distribution can be achieved by controlling the rates of nucleation and growth during synthesis.

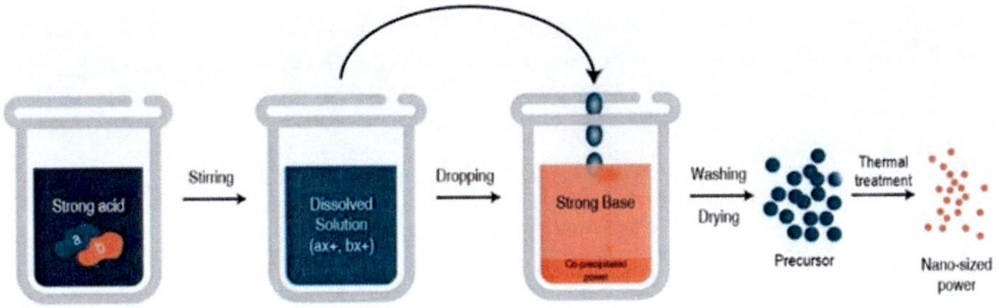

Fig. 5.8 Schematic illustration of the co-precipitation synthesis process

Co-precipitation is mainly used to obtain materials with high surface areas when compared to other synthesis methods. Magnesium ferrite nanoparticles [41] were synthesized by co-precipitation and sol–gel citrate–nitrate methods. Both samples had similar compositions and lattice parameters. However, the microstructure of the samples was different. The BET surface area of the co-precipitated sample was nearly twice greater than that of sol–gel combusted sample and consisted of elongated grains. Figure 5.9 shows the micrograph of the samples.

Xia and collaborators [42] developed a facile complex co-precipitation method for the synthesis of mesoporous Fe_3O_4 nanocages. This method differs from the

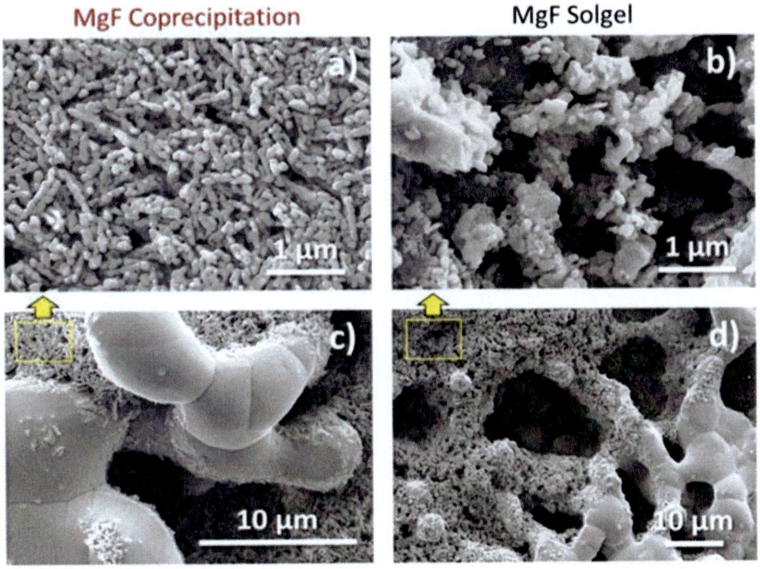

Fig. 5.9 SEM micrograph of magnesium ferrite powders prepared by co-precipitation [(**a**), (**c**)] and sol–gel combustion [(**b**), (**d**)] coated on a porous silver electrode for electrical and gas sensing measurements. Reprinted from [41], with permission from Elsevier

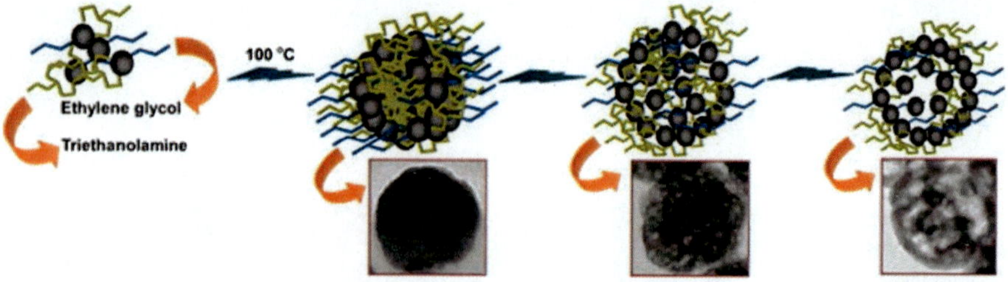

Fig. 5.10 Schematic illustration of the formation of mesoporous Fe_3O_4 nanocages. Reprinted from [42], with permission from Elsevier

conventional co-precipitation method through fast nucleation and growth. As illustrated in Fig. 5.10, the process mainly includes three steps, which can be understood as a nucleation-oriented agglomeration and Ostwald-ripening mechanism from primary small nanocrystals. This unique nanostructure is composed of very small nanoparticles with a size of 5–10 nm. In the mesoporous cage-like structure, the large specific surface area of 133 $m^2 g^{-1}$ effectively supplies the great electrolyte/electrode contact, which is beneficial for the electron transfer, lithium-ion diffusion, and the alleviation of massive volume change during the discharge/charge process., This complex co-precipitation strategy can be extended to build a variety of other mesoporous transition metal oxide nanomaterials for applications in adsorbents, catalysts, high-performance LIBs, and supercapacitors.

The co-precipitation synthesis can also be used to obtain coatings or core–shell morphologies. Fe_3O_4 nanoparticles were synthesized by Panta et al. [43] by co-precipitation of the precursors ferrous chloride tetrahydrate ($FeCl_2.4H_2O$) and ferric chloride hexahydrate ($FeCl_3.6H_2O$). Raw materials were deoxygenated with dry argon under vigorous stirring under heating at 80°C. Once the temperature of 80°C was reached, 100 ml of tetramethylammonium hydroxide were added dropwise until the solution reached pH 11.

5.8 Electrochemical

In this process, the ion source is the scarification of anode electrodes through the oxidation process. Once the anode is oxidized and released into the solution, the ions are interacting in between the electrode–electrolyte interface and form the particles. The resulting particles have a purity and size control from the current modify or distance between the electrodes. By increasing the current density or by decreasing the distance between electrodes, there is an increase in particle size. This is because the closer the electrodes, the faster the reduction/oxidation occurs and release more releasing more oxidized metal ions into the solution and, which in turn results in a fast rate of particle formation. In general, electrochemical synthesis methods are simple, inexpensive, environmentally friendly, and cost-effective. This synthesis method can be used to remove heavy metal ions from liquid waste. For example, Fe^{2+} can be generated through the electrolysis of iron anode. Moreover, part of Fe^{2+} ions could be oxidized into Fe^{3+} by controlling electrolysis conditions, so as to remove the heavy metal ions in liquid waste in the process of forming ferrites. Compared to chemical synthesis of ferrite, this method not only avoids adding the raw materials Fe^{2+} and Fe^{3+} but also has some advantages for high salinity liquid waste in the field of treatment of cobalt-containing liquid radioactive waste. Huang and collaborators [44] evaluated an innovative technological process of in situ electrochemical synthesis of ferrite to remove the Co^{2+} ions from the simulated liquid radioactive waste in a nuclear power plant. They investigated the effects on various reaction conditions (including current density, reaction time, reaction temperature, initial pH value, and boric acid concentration) and the type of power supply and discussed the reaction mechanism to provide technical support for the purification of liquid radioactive waste. Under the optimal process conditions, the current density was 40 mA/cm^2, the reaction time was 50 minutes, the reaction temperature was 60°C, the initial pH value was 8.5, the concentration of Co^{2+} was 5 mg/L, and boric acid was 1000 mg/L in the effluent. The concentration of Co^{2+} in the liquid medium was 0.447 μg/L, and the removal rate of Co^{2+} was more than 99.99%, decontamination factor (Co^{2+}) > 1.00×10^4. The best removal efficiency was achieved when the cycle for reversal of polarity was of 10 minutes. They reported that the combination of the pulse power supply with reversal polarity is better for the treatment of the simulated cobalt-containing liquid radioactive waste. This conclusion is supported by the fact that, the iron anode lost electrons and was oxidized to Fe^{2+} at the beginning of the reaction. Under the action of electrolysis and dissolved oxygen in the water, most Fe^{2+} was oxidized to Fe^{3+}. $Fe(OH)_3$ formed during the process is dehydrated to form ferrite at high pH at the end of the reaction. Co^{2+} easily forms $CoFe_2O_4$, which is also magnetic and is effectively removed. In conclusion, they perceived that the method can be effective in the removal of cobalt from liquid radioactive waste by in situ electrochemical syntheses of ferrite. It has potential applications in the actual disposal of nuclear waste in near future because of its low cost and high removal rate.

Ferrites generated by electrochemical synthesis can be used to decontaminate pollutants in the same step. A research project developed an electrocoagulation process to remove boron, in which the types of working anodes (Fe, Ni) and electrolytes (Ni, Co, Cu salts, and NaCl) were investigated to produce magnetically separable spinel ferrites [45]. Figure 5.11 shows the schematic illustration of the electrochemical synthesis process developed by Widhiastuti and collaborators.

5.9 Hydrothermal

In the hydrothermal synthesis technique, soluble salts containing divalent and trivalent transition metal ions are dissolved separately and posteriorly mixed. This can be followed by the dropwise addition of organic solvents such as ethylene glycol or ethanol to the mixtures of aqueous solutions in order to homogenize the solution.

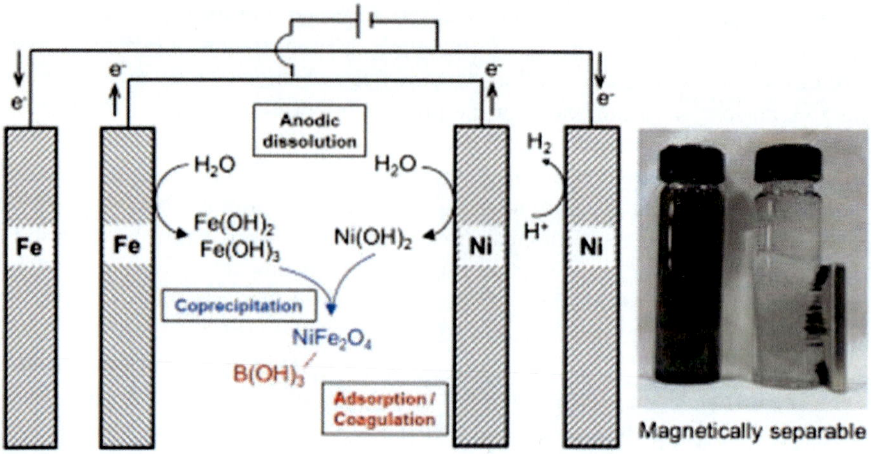

Fig. 5.11 Schematic illustration of the electrochemical synthesis process. Reprinted from [45], with permission from Elsevier

Thereafter, the solution is heated under high pressure in an autoclave. The heating temperature and time are dependent on the type of the target ferrite or particle size to be achieved. The hydrothermal method is among one of the most promising synthesis methods for the large-scale production of ferrites. Formation of good quality and high crystallinity of particles or nanoparticles with controlled size and size distribution can be achieved by selecting the appropriate mixture of solvents and varying parameters such as temperature, pressure, and reaction time. In addition, the hydrothermal method does not require extremely high processing temperature or sophisticated processing routes. For instance, crystalline ferrite nanoparticles can be prepared at temperatures as low as ~ 150°C [46, 47]. Moreover, the hydrothermal process is environmentally friendly since it does not require any kind of organic solvents or post-treatment steps such as posterior thermal treatment, which is cost-saving as well. Figure 5.12 illustrates the hydrothermal synthesis method.

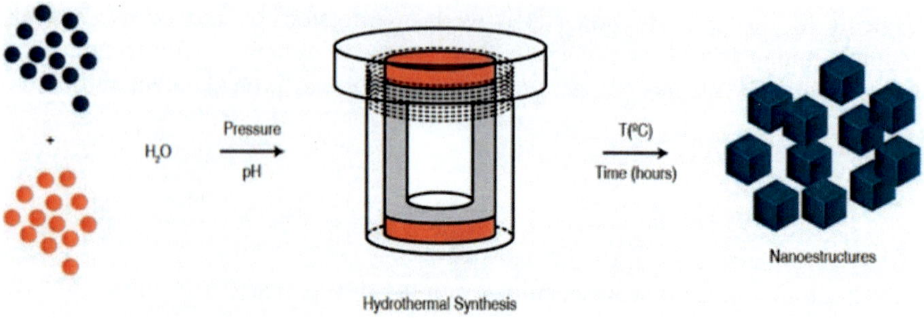

Fig. 5.12 Illustration of the hydrothermal synthesis method

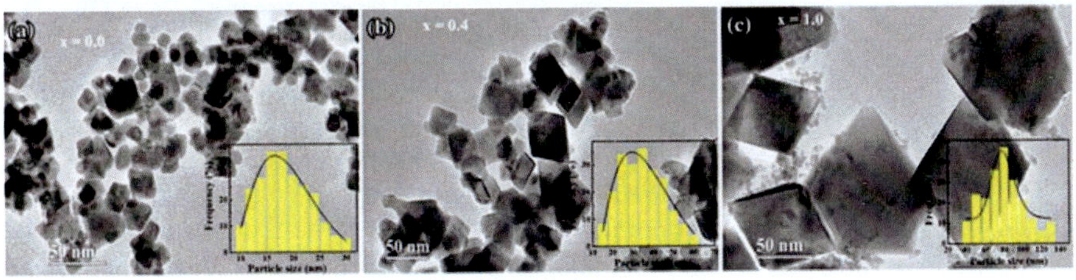

Fig. 5.13 TEM microscopy of $Co_{1-x}Ni_xFe_2O_4$ nanoparticles with representative composition x = 0.0, 0.4, and 1. Reprinted from [48], with permission from Elsevier

Melo and collaborators [48] investigated the effect of substitution with Ni^{2+} ions on the structural, optical, and magnetic properties of nickel-doped cobalt ferrite $Co_{1-x}Ni_xFe_2O_4$ ($0 \leq x \leq 1$) nanoparticles with nominal composition x = 0 – 1.0, synthesized by the hydrothermal method using nitrates of respective metals and sodium hydroxide. They obtained ferrites with nanostructured character in all samples, being roughly spherical and diamond-shaped for x = 0 and x = 1, respectively, while samples with an intermediate composition exhibited both morphologies, as shown in Fig. 5.13.

They also evaluated the cation distribution of the samples. The analyses suggested that Ni^{2+} ions substituted Co^{2+} ions in the octahedral B-sites. This effect can be observed in the optical band gap, for example. The band-gap energy values decreased with increasing Ni^{2+} ion (x) concentration. The presence of nickel in the cobalt ferrite structure affected the magnetic properties as well. For instance, the saturation magnetization, Ms, and remanent magnetization, Mr, decreased from 369 to 256 emu cm^{-3} and 131–45 emu cm^{-3} for x = 0 and x = 1, respectively.

On the other hand, the coercivity increased from 890 to 1590 Oe for x = 0 and x = 0.6, and sharply dropped to 50 Oe for x = 1. The enhanced coercivity occurred due to the particle size, defects, and residual strain, which may act as pinning centers. The magnetic results indicate that Ni-doped cobalt ferrite nanoparticles have potential in several technological applications such as magnetic sensors, spintronics, actuators, and radar absorbing materials.

5.10 Solvothermal

The solvothermal synthesis method is very similar to the hydrothermal synthesis method, but, in this case, non-aqueous solvents can be used. These solvents are used because they may aid in the precise control over the size distribution, shape, and crystalline phases. They can also greatly increase the reactivity and solubility of reactants. These physical characteristics can be altered by changing certain experimental parameters, such as the reaction temperature, reaction time, solvent, surfactant, and precursors. Usually, the solvothermal synthesis method is useful where the synthesis of ferrites with improved physical and chemical properties applicable to both industrial and biomedical areas is required. In addition to its simplicity in terms of its reaction, solvothermal synthesis is used in scalable production of particles with controlled size and shape.

The mechanism for the formation of ferrite particles can be divided into two stages: first, nucleation from the atoms formed by the reduction of metal salts, and second, the growth of nuclei into nanospheres through a process of atomic growth.

Many syntheses by solvothermal method use ethylene glycol as an additive of the process. Ethylene glycol can play multiple roles in the formation of nanocrystalline nickel ferrite particles, serving not only as solvent, reducing agent, and complexing agents for the reactants but also as stabilizing agents for the resulting nanocrystals.

Size-controlled nickel ferrite nanoparticles could be synthesized by Wang et al. [39] via a solvothermal method using ethylene glycol as solvent and NaAc as electrostatic stabilizers. This approach is a simple one-step method with a comparatively high yield. The size of the nanoparticles can be controlled from 6 to 170 nm by adjusting the experimental parameters. For example, the size of the nickel ferrite nanoparticles increases as the reaction time is prolonged and decreases gradually as the concentration of the reactants increases. On the other hand, the amount of NaAc has profound influence on the morphology and size of nickel ferrite nanoparticles. It is clear that the coercivity values decrease as the size of nickel ferrite nanoparticles decrease while the magnetization values have no similar trends, which may be attributed to the fact that the magnetic saturation value is not only affected by the size but also by the morphology.

The use of additives can change the morphologies and consequently the properties of ferrites. Zhang et al. [49] evaluated the synthesis of mesoporous manganese ferrite ($MnFe_2O_4$) microspheres composed of nanoparticles (10–30 nm). These microspheres are in the range of 100–300 nm in size and have a surface area of 60.2–86.8 $m^2.g^{-1}$, which can be tuned by varying the added amounts of CH_3COOK. The formation process of mesoporous $MnFe_2O_4$ microspheres is shown in Fig. 5.14. The authors say that firstly the Mn^{2+} and Fe^{3+} ions nucleate under solvothermal conditions

Fig. 5.14 Illustration of the formation process of mesoporous $MnFe_2O_4$ microspheres. Reprinted from [49], with permission from Elsevier

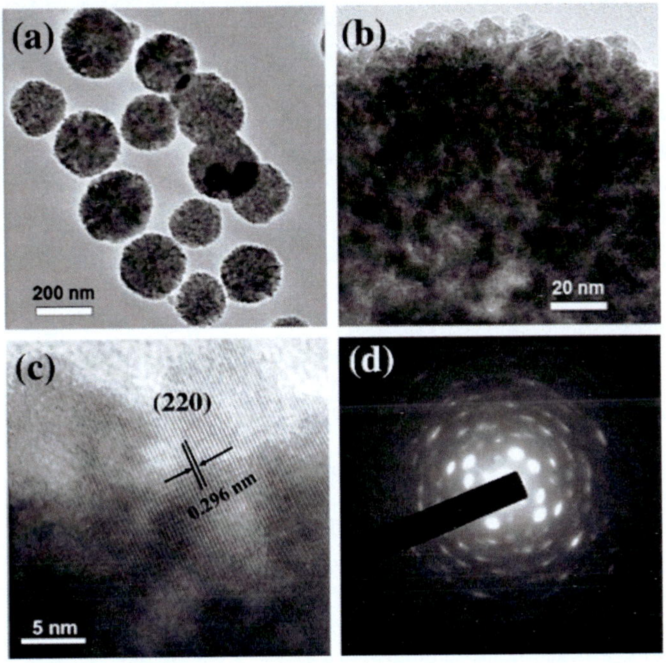

Fig. 5.15 TEM image of the mesoporous structure obtained. Reprinted from [49], with permission from Elsevier

via the reaction with water generated from metal precursors to form nanosized crystalline $MnFe_2O_4$. In the second stage, these small nanoparticles are self-assembled into large secondary particles by ethylene glycol and $CH_3COOC_2H_5$ absorbed on the surface of the nanosized crystalline particles.

Figure 5.15 shows the TEM image of the obtained mesoporous structure, revealing that these microspheres are composed of nanoparticles in the range of 10–30 nm. The TEM image in Fig. 5.3c further confirms the high crystallinity of these nanoparticles and the discontinuous diffraction rings, suggesting that $MnFe_2O_{4\text{-}2}$ microspheres are polycrystalline.

5.11 Microwave

Most synthesis methods often require long times, multiple reaction steps, precise pH control, special working conditions, and the addition of stabilizing or chelating agents. To overcome these shortcomings, microwave-assisted hydrothermal synthesis can be explored for the preparation of ferrites. Microwaves are the electromagnetic waves with the frequency between 300 MHz and 300 GHz and the wavelength in the range from 1 to 1 mm [50]. During the process of microwave

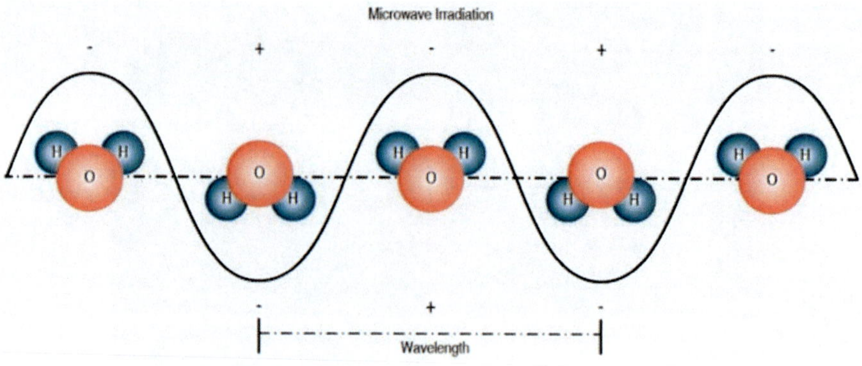

Fig. 5.16 Mechanism involved in the microwave synthesis

heating, the reaction depends on the interaction between microwave and materials. In an electromagnetic field, the particles of materials can produce polarization, such as electron polarization, atom polarization, orientation polarization, and space charge polarization. Microwave-assisted hydrothermal synthesis can produce materials with similar characteristics to those obtained by conventional hydrothermal synthesis (or solvothermal synthesis) but in a much shorter time. This method offers a high heating rate, which leads to a better precursor dissolution [51]. In a microwave system, heating arises from either dipole rotation or ion migration induced by the microwave field, which subsequently generates a better precursor dissolution, a rapid, simultaneous and homogeneous nucleation, fast supersaturation, and short crystallization time. This method allows for greater control of the products' microstructural characteristics and final properties. Figure 5.16 shows the mechanism involved in the microwave synthesis.

Compared with conventional methods, microwave irradiation has the advantages of rapid volumetric heating, high reaction rates, very short reaction times, production of uniform small particles with a narrow particle size distribution and high purity, great selectivity, and high product yield. Typically, the temperature employed in the microwave-assisted synthesis ranges from 100 to 200°C. Any vapor produced during heating time can be eliminated via an exhaust drain connected to the Teflon vessel.

In general, the microwave-assisted synthesis route offers an opportunity to the synthesis of ferrites on a large scale but with a low yield when compared to co-precipitation, hydrothermal or thermal decomposition methods. The method has been used for the synthesis of a number of ferrites, for instance, Fe_3O_4 and $CoFe_2O_4$ [52, 53]

Farinas and collaborators [53] obtained nanocrystalline cobalt ferrite powders (5–10 nm) by microwave-assisted synthesis from commercially available $Fe(NO_3)_3{\bullet}9H_2O$, $Co(NO_3)_2{\bullet}6H_2O$, ammonium hydroxide, and tetrapropylammonium hydroxide (TPAH). The stoichiometry of the synthesized $CoFe_2O_4$ powders strongly depends on both the reagents used for precipitation (NH_4OH or TPAH) and the temperature of synthesis (heating rate and dwell time). Although both precipitating reagents lead to very similar nanopowders in morphology and size, the use of TPAH instead of NH_4OH allows a dramatic reduction in the temperature of synthesis from 190°C to only 130°C. Despite this low temperature, a high degree of crystallinity can be achieved by control of synthesis parameters.

The magnetic properties strongly depend on the final stoichiometry of the material, being optimal (i.e., maximum Ms) for the nominal $CoFe_2O_4$ composition. For some synthesis conditions, a secondary magnetic phase is observed (possibly Fe_3O_4). Such a composition can be tuned via the precursor used as a base and the annealing temperature. When TPAH is used as a base for the precipitation, the net magnetization is larger as compared with the NH_4OH counterpart prepared under the same conditions, but the resulting material is magnetically non-homogeneous. Therefore, the synthesis with NH_4OH at 190°C seems the most appropriate to obtain a homogeneous material with improved magnetic properties in terms of saturation magnetization, remanence, and coercivity.

5.12 Sonochemical

A relatively new route for the synthesis of ferrite is by using ultrasound, in which the physical and chemical effects are induced by cavitation (i.e., nucleation, growth, and the transient collapse of tiny gas bubbles driven by pressure variation created by ultrasound) are responsible for the reactions and synthesis. Upon collapse, an extremely high energy is released to heat the content of the bubble. This generates a transient localized hot spot with an effective temperature and pressure inside the bubbles. Figure 5.17 shows the mechanism involved in the sonochemical synthesis.

These extraordinary conditions permit access to a range of chemical reactions, which are normally inaccessible via other methods. The composition of the particles expected to be synthesized using the sonochemical method is identical to the composition of the vapor in the bubbles; this assists in controlling the purity of produced ferrite particles. The sonochemical method is, especially, important in the uniformity of mixing, reduction of crystal growth, and control over the particle size distribution,

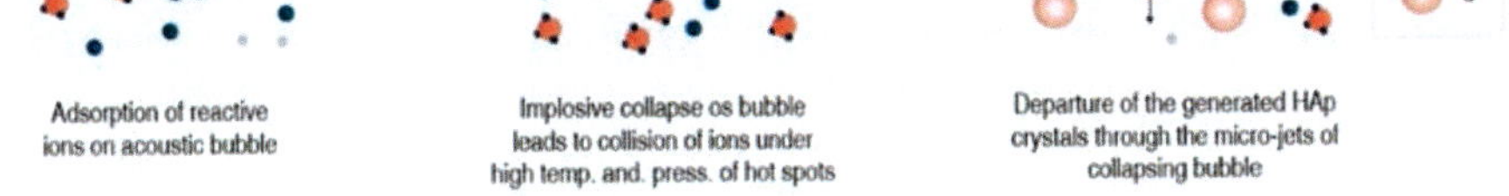

Fig. 5.17 Mechanism involved in the sonochemical synthesis

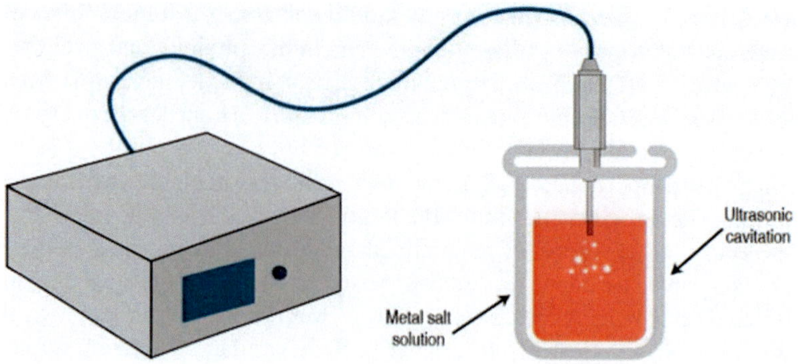

Fig. 5.18 Sonochemical synthesis process

but less important for the synthesis of particles with controllable shapes and dispersity. When sonochemical synthesis is used, temperature and ultrasonic intensity are the major determining factors that influence particle size. Figure 5.18 depicts the sonochemical synthesis process.

Choudhury and collaborators [54] have attempted to determine the physical mechanism of sonochemical synthesis of zinc ferrite by identifying the exact role played by the chemical and physical effects produced due to cavitation events in the reaction mixture. They have also tried to find out the relative contribution of each of these effects to the overall synthesis chemistry. The major conclusion of the study is that transient cavitation in the reaction medium provides the in situ calcination of oxide particles to yield ferrite. This calcination could be a result of highly energetic collision between oxide particles due to shock waves generated by cavitation bubbles, or the heating-up of a thin liquid shell containing metal oxide particles surrounding the bubble interface during transient collapse. The major sonochemical effect is the formation of H_2O_2 that helps converting hydroxides into oxides, but in the absence of micro-turbulence, the sonochemical effect alone is not able to yield the ferrite. An added advantage of the sonochemical synthesis of ferrite through in situ micro-calcination is the smaller size of ferrite crystal, as compared to external calcination, which could yield agglomeration.

This synthesis is also an easy method for doping of ferrites. Yadav et al. [55] evaluated this method for Gd^{3+} doped cobalt ferrite nanoparticles. Their study confirmed that the sonochemical synthesis is favorable to form single spinel gadolinium-doped cobalt ferrite nanoparticles with concentrations of gadolinium up to $x = 0.20$. The average crystallite size decreases with the substitution of Gd^{3+} in cobalt ferrite nanoparticles. Moreover, the addition of Gd^{3+} in cobalt ferrite nanoparticles influences cation distribution at the tetrahedral and octahedral site. The electrical properties showed that Gd^{3+} doped cobalt ferrite exhibits enhanced dielectric constant and AC conductivity. Magnetic properties can also be tuned with ultrasonic irradiation and gadolinium substitution in cobalt ferrite nanoparticles synthesized by a sonochemical method.

5.13 Conclusions and Perspectives

In recent years, spinels have been obtained by a variety of chemical and physical methods, as previously described. Conventional methods for the synthesis of these materials are developing rapidly, focusing on several aspects such as mild synthesis condition, accurate control of size and morphology and so on. Moreover, the study and development of synthesis methods contribute to the better understanding of the relationship structure/properties of ferrites. Each current synthetic method has its particular characteristics. However, it is promising to compare or combine different synthesis methods for a deeper understanding of the modification and improvement of the intrinsic properties of ferrites.

References

1. Javed A, Szumiata T, Sarwar A, Fatima T (2019) Structure and Mössbauer spectroscopy studies of Ni0.5Zn0.5NdxFe2-xO4 ($0.00 \le x \le 0.10$) ferrites. Mater Chem Phys 221:99–107. https://doi.org/10.1016/J.MATCHEMPHYS.2018.09.042
2. Zhang Z, Yao G, Zhang X et al (2015) Synthesis and characterization of nickel ferrite nanoparticles via planetary ball milling assisted solid-state reaction. Ceram Int 41:4523–4530. https://doi.org/10.1016/j.ceramint.2014.11.147
3. Zhang Z, Liu Y, Yao G et al (2013) Synthesis and characterization of NiFe $_2$ O $_4$ nanoparticles via solid-state reaction. Int J Appl Ceram Technol 10:142–149. https://doi.org/10.1111/j.1744-7402.2011.02719.x
4. Kopp Alves A, Bergmann CP, Berutti FA (2013) Novel synthesis and characterization of nanostructured materials. https://doi.org/10.1007/978-3-642-41275-2
5. Marinca T, Chicinas I, Isnard O, Pop V (2011) Structural and magnetic properties of nanocrystalline ZnFe2O4 powder synthesized by reactive ball milling. Optoelectron Adv Mater—Rapid Commun 5:39–43
6. Marinca TF, Chicinaş I, Isnard O, Neamţu BV (2016) Nanocrystalline/nanosized manganese substituted nickel ferrites—Ni1−xMnxFe2O4 obtained by ceramic-mechanical milling route. Ceram Int 42:4754–4763. https://doi.org/10.1016/J.CERAMINT.2015.11.155
7. Chicinas I, Marinca TF, Neamtu BV et al (2014) Synthesis, structural, and magnetic properties of nanocrystalline/nanosized manganese-nickel ferrite-Mn0.5Ni0.5Fe2O4. IEEE Trans Magn 50:1–4. https://doi.org/10.1109/TMAG.2013.2285246
8. Marinca TF, Chicinaş I, Isnard O, Popescu V (2012) Nanocrystalline/Nanosized Ni $_{1-\gamma}$ Fe $_{2+\gamma}$ O $_4$ Ferrite Obtained by Contamination with Fe During Milling of NiO-Fe $_2$ O $_3$ Mixture. Structural and Magnetic Characterization. J Am Ceram Soc 96:n/a-n/a. https://doi.org/10.1111/jace.12043
9. Lysenko E, Nikolaev E, Vlasov V, Surzhikov A (2018) Microstructure and reactivity of Fe2O3-Li2CO3-ZnO ferrite system ball-milled in a planetary mill. Thermochim Acta 664:100–107. https://doi.org/10.1016/J.TCA.2018.04.015
10. Xiao J, Liu P, Wang CX, Yang GW (2017) External field-assisted laser ablation in liquid: An efficient strategy for nanocrystal synthesis and nanostructure assembly. Prog Mater Sci 87:140–220. https://doi.org/10.1016/J.PMATSCI.2017.02.004
11. Svetlichnyi VA, Shabalina AV, Lapin IN et al (2018) Study of iron oxide magnetic nanoparticles obtained via pulsed laser ablation of iron in air. Appl Surf Sci 462:226–236. https://doi.org/10.1016/j.apsusc.2018.08.116
12. Bulai G, Gurlui S, Caltun OF, Focsa C (2015) Pure and rare earth doped cobalt ferrite laser ablation: space and time resolved optical emission spectroscopy

13. Shukla A, Singh SC, Pandey BK et al (2015) Liquid-assisted pulsed laser ablation synthesized titanium ferrite nanoparticles: structural, optical and magnetic properties. Adv Mater Lett 6:1066–1072
14. Patil KC, Aruna ST, Mimani T (2002) Combustion synthesis: an update. Curr Opin Solid State Mater Sci 6:507–512. https://doi.org/10.1016/S1359-0286(02)00123-7
15. Chavarriaga EA, Lopera AA, Wermuth TB et al (2018) Synthesis by solution combustion and optical characterization of violet NASICON-type Mg0.45Co0.05Ti2(PO4)3 pigment. Dye Pigment 157:1–10. https://doi.org/10.1016/J.DYEPIG.2018.04.041
16. Kopp A, Bergmann CP, Berutti FA (2013) Novel synthesis and characterization of nanostructured materials. Springer, Berlin Heidelberg, Berlin, Heidelberg
17. Zampiva RYS, Kaufmann CG, Venturini J, et al (2021) Role of the fuel stoichiometry and post-treatment temperature on the spinel inversion and magnetic properties of NiFe2O4 nanoparticles produced by solution combustion synthesis. Mater Res Bull 138:111238. https://doi.org/10.1016/j.materresbull.2021.111238
18. Deganello F, Tyagi AK (2018) Solution combustion synthesis, energy and environment: best parameters for better materials. Prog. Cryst. Growth Charact. Mater.
19. Toniolo J, Takimi AS, Andrade MJ et al (2007) Synthesis by the solution combustion process and magnetic properties of iron oxide (Fe3O4and α-Fe2O3) particles. J Mater Sci 42:4785–4791. https://doi.org/10.1007/s10853-006-0763-7
20. Da Dalt S, Takimi AS, Sousa VC, Bergmann CP (2009) Magnetic and structural characterization of nanostructured MgFe2O4 synthesized by combustion reaction. Part Sci Technol 27:519–527. https://doi.org/10.1080/02726350903328662
21. Da Dalt S, Takimi AS, Volkmer TM et al (2011) Magnetic and Mössbauer behavior of the nanostructured MgFe2O4 spinel obtained at low temperature. Powder Technol 210:103–108. https://doi.org/10.1016/j.powtec.2011.03.001
22. González-Cortés SL, Imbert FE (2013) Fundamentals, properties and applications of solid catalysts prepared by solution combustion synthesis (SCS). Appl Catal A Gen 452:117–131. https://doi.org/10.1016/J.APCATA.2012.11.024
23. Young Sun Zampiva R, Gabriel Kaufmann Jr C, Kopp Alves A influence of the fuel composition and the fuel/oxidizer ratio on the combustion solution synthesis of MgFe2O4 Catalyst Nanoparticles. https://doi.org/10.5937/fmet1802157Z
24. Jain SR, Adiga KC, Pai Verneker VR (1981) A new approach to thermochemical calculations of condensed fuel-oxidizer mixtures. Combust Flame 40:71–79. https://doi.org/10.1016/0010-2180(81)90111-5
25. Venturini J, Zampiva RYS, Arcaro S, Bergmann CP (2018) Sol-gel synthesis of substoichiometric cobalt ferrite (CoFe2O4) spinels: influence of additives on their stoichiometry and magnetic properties. Ceram Int 44:12381–12388. https://doi.org/10.1016/J.CERAMINT.2018.04.026
26. Sriondee M, Dungsuwan W, Thountom S (2018) Synthesis and characterization of Bi0.5(Na1-xKx)0.5TiO3 powders by sol–gel combustion method with glycine fuel. Ceram Int 44:S168–S171. https://doi.org/10.1016/j.ceramint.2018.08.122
27. Vara BBVS, Ramesh KV, Srinivas A (2018) Structural and magnetic properties of nanocrystalline nickel ferrite (NiFe2O4) synthesized in sol-gel and combustion routes. Solid State Sci 86:86–97. https://doi.org/10.1016/j.solidstatesciences.2018.10.008
28. Ansari F, Sobhani A, Salavati-Niasari M (2016) Facile synthesis, characterization and magnetic property of CuFe12O19 nanostructures via a sol-gel auto-combustion process. J Magn Magn Mater 401:362–369. https://doi.org/10.1016/j.jmmm.2015.10.049
29. Erhardt CS, Caldeira LE, Venturini J et al (2020) Sucrose as a sol-gel synthesis additive for tuning spinel inversion and improving the magnetic properties of CoFe2O4 nanoparticles. Ceram Int 46:12759–12766. https://doi.org/10.1016/j.ceramint.2020.02.044
30. Erhardt CS, Caldeira LE, Venturini J et al (2020) Sucrose as a sol-gel synthesis additive for tuning spinel inversion and improving the magnetic properties of CoFe2O4 nanoparticles. Ceram Int. https://doi.org/10.1016/j.ceramint.2020.02.044

31. Venturini J, Zampiva RYS, Piva DH et al (2018) Conductivity dynamics of metallic-to-insulator transition near room temperature in normal spinel $CoFe_2O_4$ nanoparticles. J Mater Chem C 6:4720–4726. https://doi.org/10.1039/C8TC00099A
32. Mariosi FR, Venturini J, da Cas VA, Bergmann CP (2020) Lanthanum-doped spinel cobalt ferrite (CoFe2O4) nanoparticles for environmental applications. Ceram Int 46:2772–2779. https://doi.org/10.1016/j.ceramint.2019.09.266
33. Vitor PAM, Venturini J, da Cunha JBM, Bergmann CP (2021) The influence of cation distribution on the magnetic properties of mixed Co1-yNiyFe2O4 nanoferrites produced by the sol-gel method. J Alloys Compd. https://doi.org/10.1016/j.jallcom.2020.156799
34. Venturini J, Wermuth TB, Machado MC et al (2019) The influence of solvent composition in the sol-gel synthesis of cobalt ferrite (CoFe2O4): A route to tuning its magnetic and mechanical properties. J Eur Ceram Soc. https://doi.org/10.1016/J.JEURCERAMSOC.2019.01.030
35. Kakade SG, Ma Y-R, Devan RS et al (2016) Dielectric, complex impedance, and electrical transport properties of Erbium (Er^{3+}) Ion-substituted nanocrystalline, Cobalt-Rich Ferrite (Co $_{1.1}$ Fe $_{1.9-x}$ Er $_x$ O $_4$). J Phys Chem C 120:5682–5693. https://doi.org/10.1021/acs.jpcc.5b11188
36. Ahmad I, Abbas T, Ziya AB, Maqsood A (2014) Structural and magnetic properties of erbium doped nanocrystalline Li-Ni ferrites. Ceram Int 40:7941–7945. https://doi.org/10.1016/j.ceramint.2013.12.142
37. Peng Z, Fu X, Ge H et al (2011) Effect of Pr3+ doping on magnetic and dielectric properties of Ni–Zn ferrites by "one-step synthesis." J Magn Magn Mater 323:2513–2518. https://doi.org/10.1016/J.JMMM.2011.05.033
38. Nag S, Roychowdhury A, Das D et al (2018) Structural and magnetic properties of erbium (Er3+) doped nickel zinc ferrite prepared by sol-gel auto-combustion method. J Magn Magn Mater 466:172–179. https://doi.org/10.1016/J.JMMM.2018.06.084
39. Lubini M, Chinarro E, Moreno B et al (2016) Electrical Properties of $La_{0.6}Sr_{0.4}Co_{1-y}Fe_yO_3$ ($y = 0.2$–1.0) fibers obtained by electrospinning. J Phys Chem C 120:64–69. https://doi.org/10.1021/acs.jpcc.5b09696
40. Lubini M, Chinarro E, Moreno B et al (2017) Electrochemical characteristics of La0.6Sr0.4Co1−yFeyO3 (y=0.2–1.0) fiber cathodes. Ceram Int 43:8715–8720. https://doi.org/10.1016/J.CERAMINT.2017.04.002
41. Sumangala TP, Pasquet I, Presmanes L et al (2018) Effect of synthesis method and morphology on the enhanced CO2sensing properties of magnesium ferrite MgFe2O4. Ceram Int 44:18578–18584. https://doi.org/10.1016/j.ceramint.2018.07.082
42. Xia T, Xu X, Wang J et al (2015) Facile complex-coprecipitation synthesis of mesoporous Fe3O4 nanocages and their high lithium storage capacity as anode material for lithium-ion batteries. Electrochim Acta 160:114–122. https://doi.org/10.1016/J.ELECTACTA.2015.02.017
43. Panta PC, Romero RPP, Forte SK, Bergmann CP (2014) Magnetic synthesis and characterization of superparamagnetic nanoparticles iron oxide stabilized with dextran. In: Ceramic Transactions. pp 137–145
44. Huang G, Dou P, Zhang Z, Yan J (2018) Removal of cobalt from liquid radioactive waste by in situ electrochemical synthesis of ferrite. J Radioanal Nucl Chem 316:61–70. https://doi.org/10.1007/s10967-018-5766-8
45. Widhiastuti F, Lin JY, Shih YJ, Huang YH (2018) Electrocoagulation of boron by electrochemically co-precipitated spinel ferrites. Chem Eng J 350:893–901. https://doi.org/10.1016/j.cej.2018.06.041
46. Phumying S, Labuayai S, Swatsitang E et al (2013) Nanocrystalline spinel ferrite (MFe2O4, M = Ni Co, Mn, Mg, Zn) powders prepared by a simple aloe vera plant-extracted solution hydrothermal route. Mater Res Bull 48:2060–2065. https://doi.org/10.1016/J.MATERRESBULL.2013.02.042
47. Cheng Y, Zheng Y, Wang Y et al (2005) Synthesis and magnetic properties of nickel ferrite nano-octahedra. J Solid State Chem 178:2394–2397. https://doi.org/10.1016/J.JSSC.2005.05.006

48. Melo RS, Banerjee P, Franco A (2018) Hydrothermal synthesis of nickel doped cobalt ferrite nanoparticles: optical and magnetic properties. J Mater Sci Mater Electron 29:14657–14667. https://doi.org/10.1007/s10854-018-9602-2
49. Zhang Z, Wang Y, Tan Q et al (2013) Facile solvothermal synthesis of mesoporous manganese ferrite (MnFe2O4) microspheres as anode materials for lithium-ion batteries. J Colloid Interface Sci 398:185–192. https://doi.org/10.1016/j.jcis.2013.01.067
50. Meng L-Y, Wang B, Ma M-G, Lin K-L (2016) The progress of microwave-assisted hydrothermal method in the synthesis of functional nanomaterials. Mater Today Chem 1–2:63–83. https://doi.org/10.1016/J.MTCHEM.2016.11.003
51. Wermuth TB, Baibich MN, Ribeiro TMH, Bergmann CP (2018) The rapid synthesis of nanostructured orthorhombic KNbO3 particles by a microwave-assisted hydrothermal method and their characterization. Ceram Int 44:4758–4765. https://doi.org/10.1016/j.ceramint.2017.12.060
52. Manikandan A, Sridhar R, Arul Antony S, Ramakrishna S (2014) A simple aloe vera plant-extracted microwave and conventional combustion synthesis: Morphological, optical, magnetic and catalytic properties of CoFe2O4nanostructures. J Mol Struct 1076:188–200. https://doi.org/10.1016/j.molstruc.2014.07.054
53. Fariñas JC, Moreno R, Pérez A et al (2018) Microwave-assisted solution synthesis, microwave sintering and magnetic properties of cobalt ferrite. J Eur Ceram Soc 38:2360–2368. https://doi.org/10.1016/J.JEURCERAMSOC.2017.12.052
54. Choudhury HA, Choudhary A, Sivakumar M, Moholkar VS (2013) Mechanistic investigation of the sonochemical synthesis of zinc ferrite. Ultrason Sonochem 20:294–302. https://doi.org/10.1016/j.ultsonch.2012.06.006
55. Yadav RS, Kuřitka I, Vilcakova J et al (2018) Sonochemical synthesis of Gd3+ doped CoFe2O4 spinel ferrite nanoparticles and its physical properties. Ultrason Sonochem 40:773–783. https://doi.org/10.1016/J.ULTSONCH.2017.08.024

Chapter 6
Processing

Ceramic processing can be defined as the set of operations by which, from one or more starting materials, parts with desired shape and structure are obtained. The advent of materials science and the development of new technologies allowed ceramic products to be inserted in the most diverse sectors of the industry. This chapter covers the fundamental concepts involved in the various stages of ferrite processing, from the dispersion of particles to the consolidation of a bulk or thin film.

6.1 Introduction

Although several spinel ferrites applications are concentrated in the synthesized powders, many applications also use these promising materials in bulk, thin films, thick layers, multilayer materials, etc. In all these cases, the powders previously synthesized go through a crucial stage: processing.

The processing stage is extensive and may even include the synthesis stage, as discussed in the previous section (Chapter 5), especially when discussing colloidal processing. In this chapter, we start from the principle that we already have the synthesized powder, and now we will consolidate it with an appropriate microstructure for its application. The key to understanding ferrite processing—as that of all ceramic materials—is to recognizing the interdependence of the different stages of the process and being able to obtain the appropriate relationships. Correlations between processing, structure, properties, and applications can be found through appropriate characterizations, as showed in Fig. 6.1.

S. Arcaro and J. Venturini, *Modern Ferrites in Engineering*, Topics in Mining, Metallurgy and Materials Engineering, https://doi.org/10.1007/978-3-030-78988-6_6

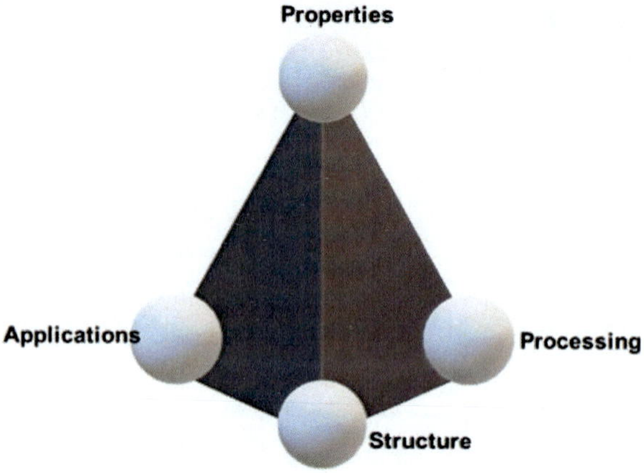

Fig. 6.1 Connections between processing, structure, properties, and application variables, promoted through appropriate characterization

6.2 Ferrite Bulk

The ferrite powders must be consolidated into a bulk with a microstructure appropriate for each application. In this case, straightforwardly, requirements can be met by carefully controlling three processing steps:

1. Powder preparation;
2. shaping;
3. sintering.

These steps are closely interconnected the characteristics of the synthesized powder can strongly affect the product's quality after sintering. The remaining porosity or other powder defects can be corrected by extended times or higher temperatures in the sintering step, but usually at the cost of deterioration of different properties. Therefore, the optimal combination is a collaborative process between powder synthesis and powder control, shaping method, and sintering.

Among the characteristics of the particle systems used to obtain dense samples, it is essential to have high purity particles of the smallest possible average size. The smaller the average particle size, the higher the surface area and, consequently, the higher the energy associated with the system, and the lower the sintering temperature required to obtain materials with high relative density. The powders must have a narrow particle size distribution to avoid excessive grain growth, and the formation of agglomerates should also be avoided, as these sinter faster than the surrounding particles, generating internal cracks after sintering [1–3].

Therefore, different strategies can be used to produce bulk ferrites, either dense or porous: various powder stabilization technique, shaping, and sintering methods.

6.2.1 *Powder Preparation*

After the ferrite synthesis, powders usually have high reactivity, whether they are nanometric or micrometric. This reactivity reduces the flow of the powders when subjected to conformation processes. The homogeneity of the final material will depend on the ability of the initial particles to reorganize themselves during high-temperature sintering. It is worth noting that ferrites and all ceramics with the best properties are obtained when the compact sinters at a density close to the theoretical value, except in the case of porous materials with desired porosity. In most cases, the powder preparation or the processing itself involves some stage in a wet state, which requires an adequate dispersion of the powder in the liquid for which some additives are needed. Therefore, an understanding of colloidal processing is necessary. It is impossible to produce materials with suitable final properties by wet processing without the suspensions' control and stability.

6.2.1.1 Colloidal Processing

The class of colloidal particles comprises particles with dimensions between 1 nm and 1 μm. Particles with colloidal sizes are known to have a high surface area. From an energetic point of view, colloids can be stable (reversible) or unstable (irreversible). This last case is the one that includes ceramic materials in general. It is expected that these particles of such small size tend to clump together to reduce their surface energy [2, 4, 5]. Colloidal processing plays a determining role in many ceramic manufacturing steps. Although the colloidal size is typically <1 μm, the same approach may be successfully applied in systems with larger particle sizes. Likewise, magnetic colloids consist of macroscopically homogeneous dispersions and microscopically heterogeneous because they are composed of solid particles with diameters of the order of a few nanometers, dispersed in a liquid medium. Due to the small size of the particles, interfacial phenomena govern the global behavior of these systems.

When applying any of the colloidal processing methods, one of the most critical steps in preparing stable and homogeneous suspensions is developing versatile and robust routes to stabilize the magnetic systems. The particles dispersed in a liquid medium are in constant random motion, known as Brownian motion. This phenomenon causes a high frequency of collisions between particles. The stability of colloidal suspensions is directly dependent on the way the particles interact during the collision. The main factors determining the properties of a colloidal system are the particle size and shape, the surface properties, the interparticle interactions, and the interactions between particles and the dispersing medium, i.e., the solid–liquid interface. The colloidal approach involves manipulating and controlling those interactions to optimize the suspension properties to produce green bodies with a reduced number of heterogeneities, better microstructural control, and enhanced features [2].

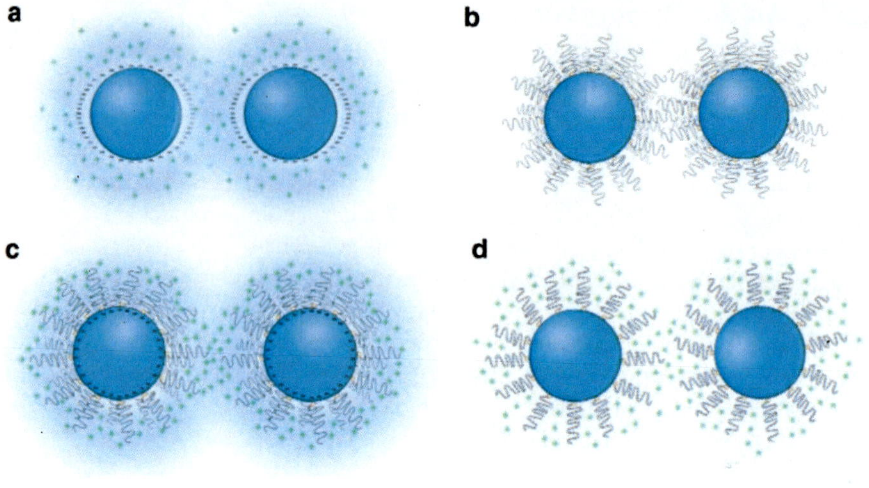

Fig. 6.2 Schematic representation of the stabilization mechanisms for nanofluids: **A** electrostatic stabilization, **B** steric stabilization, **C, D** electrosteric stabilization. Reprinted from [6], with permission from Elsevier

The suspension stabilization mechanism is based on balancing the repulsive and attractive forces: the development of electrical charges on the particle act due to the interaction of its surface with the liquid (electrostatic) medium; polymers of the long-chain hinders the approximation of the particles by the mechanical (steric) impediment; and the specific adsorption of molecules with ionizable groups or polyelectrolytes on the surface of the particles (electrostatic), in which the ions resulting from the dissociation of these ionizable groups add an electrostatic barrier to the steric effect, seen in Fig. 6.2. These phenomena change the particle/liquid interface of the system, promoting dispersion of the suspension [5].

To increase the particles' repulsive forces to avoid agglomeration, there is an electrostatic stabilization process based on the increase of the particles' surface charge. The electrostatic stabilization mechanism consists of the formation of an ion cloud around the particles. Thus, when powders are inserted in a liquid medium, a cloud of electrical charges is instantly formed around it, known as a double electrical layer, as shown in Fig. 6.3. This region is formed by the first layer of ions of opposite signs (counter-ions) that are firmly adsorbed on the surface by electrostatic attraction and a second outer layer (diffuse layer). The interface region between the two layers is called the Stern layer, where shear occurs during the liquid flow [2, 7]. This layer is because solid particles have on their surfaces many unsaturated atomic bonds capable of creating bonds with species present in the suspension, which may be hydroxonium (H_3O^+) or hydroxyl (OH^-) ions in addition to any other ion present in the medium.

The processes that take place also generate counter-ions. These counter-ions promote a linear decay of potential until reaching the Stern layer. This region is

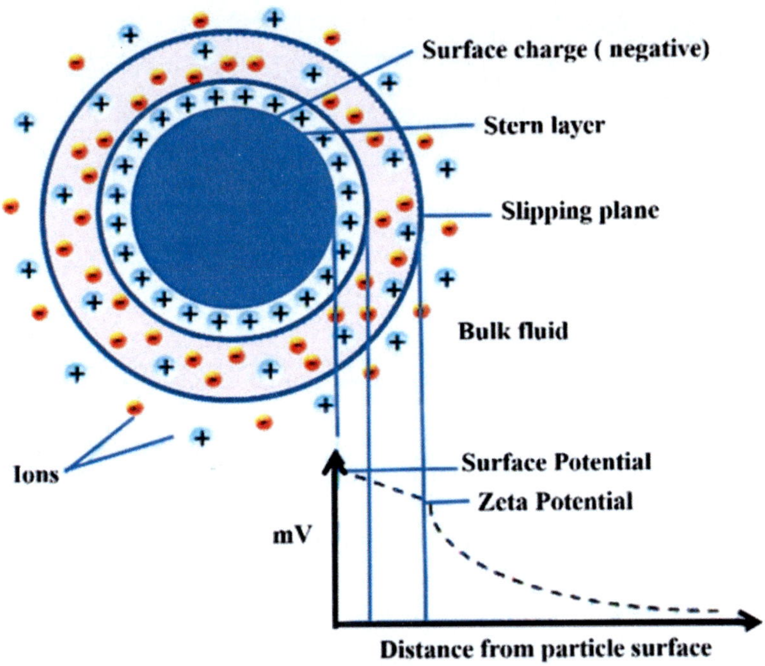

Fig. 6.3 Spatial arrangement of the ions that constitute the double electrical layer

crucial in the chemical characterization of the particles. Upon application of an electrical potential, the speed of the particles (electrokinetic mobility) increases. This value is the zeta potential of the particle and is used to evaluate the repulsive potential between the particles [2, 5]

Knowing that the zeta potential is determined indirectly from electrophoretic mobility measurements as a function of the pH of the liquid medium, then the pH value can be determined where the surface activities of positively and negatively charged ions are equal. This value is equivalent to the neutralization of positive and negative charges and is known as the zero charge point. The pH at which the particle repulsion potential is zero is known as the isoelectric point. In the absence of explicitly adsorbed ions, the zero charge point and isoelectric point values coincide [2, 8].

The increase in the distance from the zero charge point increases the condition of stability. Colloidal suspensions are considered stable if the particles' zeta potential measurements are above +30 mV to below −30 mV. Otherwise, if the pH is in the range of the zero charge point, the particles will tend to agglomerate and establish a rigid structure (gel) in the form of a flocculated suspension (Table 6.1).

The electrostatic mechanism has some limitations regarding the control of suspension stability: dependence on the pH of the suspension and the concentration of electrolytes. These variables can increase the ionic strength of the liquid and cause

Table 6.1 Isoelectric points of some ferrites

Ferrite	IEP	Reference
$CoFe_2O_4$	7.5	[9]
$NiFe_2O_4$	6.7;7;9	[9]
$FeTa_2O_6$	3.8	[10]
$Y_3Fe_5O_{12}$	5–6.8	[10]
Fe_3O_4	6	[11]
$ZnFe_2O_4$	4.6	[12]
$MgFe_2O_4$	5.2	[13]

particle agglomeration. The steric mechanism is used to solve the agglomeration. The principle is based on the adsorption of organic molecules (usually of polymeric origin), which act as agents of physical repulsion. For the polymer to be effective, there must be an attractive interaction between the particle surface and the polymer. Also, the layer thickness must be large enough not to fall within Van der Waals forces' substantial acting distance. Finally, the polymer needs to have an affinity for the liquid. Compared with the electrostatic mechanism, the steric mechanism has few advantages [2, 5].

a. Little sensitivity to electrolytes: Electrostatically stabilized suspensions behave differently if the liquid's ionic strength is modified, which in turn alters the configuration of the double electrical layer.
b. Variation of the liquid medium: Sterically stabilized suspensions are efficient in both aqueous and non-aqueous media.
c. Flocculation reversibility: Sterically stabilized suspensions are in a favorable thermodynamic condition. In contrast, electrostatically stabilized suspensions are in metastable equilibrium, and the state of particle agglomeration is that of less energy, making redispersion difficult.

Colloidal suspensions can be stabilized by the presence of an ionic cloud around the particles (electrostatic stabilization) and by the addition of polymers. The last stabilization mechanism is the combination of these two effects, known as the electrosteric stabilization mechanism. In this mechanism, specific dispersing polymers have ionizable groups that adsorb onto the particles. These polymers are called polyelectrolytes. In the electrosteric mechanism, the particle's surface charge is also influenced by the electrical charge developed by the dissociation of the electrolyte. Polyacrylates are the preferred dispersants for most situations.

Still, in the specific case of magnetic colloids, the dispersed phase comprises magnetic particles that individually present non-zero magnetic moments. The magnetic dipolar interaction between a particle and its neighbors will depend on the relative orientation of the magnetic moments, but it usually presents itself as an attractive interaction, which can induce aggregate formation.

The potential of magnetic dipolar interaction is dependent on the relative orientation between the magnetic moments. Maximum attraction occurs when the moments are aligned in the same direction. On the other hand, the magnetic dipolar interaction is considered negligible when the magnetic coupling is low enough. This can occur when the moment is tiny, or when a significant distance separates the particles, or if the thermal energy is sufficiently large. In a magnetic fluid, the distance between particles is a parameter that can be strategically adjusted through the concentration of particles in the colloid. In this context, the more diluted the system, the greater the average distance between the particles. Nevertheless, the intensity of the net magnetic moments will determine how far apart the particles must be from each other so that the magnetic coupling is effectively negligible.

In colloidal dispersions, attractive interactions (Van der Waals and magnetic dipolar) must be compensated for by repulsive interactions to prevent the aggregation process, leading to colloid coagulation. Thus, they can be stabilized with molecules that provide a physical barrier, leading to a steric-type repulsion or introducing an electrostatic barrier through a density of surface charges [14].

The analysis of the existing equilibria between the surface and the dispersion allows us to determine the dependence of the surface charge density as a function of the pH of the medium. Using simultaneous measurements of potentiometry and conductivity, it is possible to determine the surface charge density variation as a function of pH. In a neutral medium, this density is low, Coulomb repulsions between particles are ineffective, and magnetic colloids are not stable. For pH values below 3.5 and above 10.5, the nanoparticles' surface is electrically saturated, and the electrostatic interaction between particles is usually enough to stabilize the colloid.

6.2.1.2 Rheology: Basic Concepts

Rheology studies the behavior of homogeneous fluids under the action of stresses. This is a field of study of great importance in the industrial processing of particulate materials due to the need for production, quality, storage, pumping, and transportation; wherein at least in one step of the process, there was a need to disperse powders in a liquid medium [2, 4]. When carrying out a rheological study for particle dispersion, it is necessary to know the characteristics of the suspension. These characteristics facilitate the interpretation of the obtained curves, correlating results and conditions of production and processing of suspensions [15].

In colloidal processing, the function of the liquid medium in the system is to act as a particle dispersion agent to dissolve salts, polymers, and other additives within the suspension. The type of liquid (aqueous or non-aqueous) used in the colloidal suspension composition alters its consistency; therefore, the rheological behavior is also altered. Water is the most usual solvent due to its lower cost, wide availability, and absence of toxicity. Organic liquids, such as alcohol and acetone, were used when the particles that make up the suspension react with water or when the drying step is problematic [16].

In highly concentrated suspensions and the concentration of solids, the morphology of the particles and their size distribution also influence the viscosity. The flow of the suspension causes particle orientation, and the higher the aspect ratio of the particles, the greater the turbulence generated by the dissipation of energy.

Smaller particle size suspensions exhibit the highest viscosity values. This behavior is attributed to the significant interaction between double electrical layers, so that, by decreasing the particle size, the extension of the double layer is increased and the free average path between the particles decreases. Suspensions made up of particles with wide size distribution ensure better packing, allowing a reduction in viscosity, as smaller particles fill the larger particles' interstices, decreasing the solid volume. Consequently, the water layer becomes thicker and reduces the suspension's viscosity [7, 17, 18].

Concentrated suspensions have a minimum amount of stress needed to start their flow, called the yield stress. This tension arises from the presence of particles agglomerated in a way that a rigid three-dimensional structure is formed. This occurs mainly due to particles with a high surface area and/or different natures (particles with opposite charge signs) [19, 20]. Roughness is also a factor that can generate agglomeration due to the ease of interparticle contact. The interlacing of polymeric chains is another factor that affects yield stress [5].

In general, the rheological behavior of suspensions can be classified according to five models: Newtonian, pseudoplastic, dilator plastic, and Bingham plastic fluids. Isaac Newton, in 1687, was responsible for the first mathematical approach, correlating the strain rate (γ) that a fluid undergoes with the application of external stress (τ). Considering that the fluids deform continuously, it is not possible to determine an absolute strain value. Thus, the rheological characterization of suspensions comprises deformation rates as a function of time ($\dot{\gamma}$). Newton's model is explained by the representation of two parallel sheets separated at a distance dx. These blades move at different speeds "v1" and "v2" under the action of a given force F [4].

The volume contained in these blades undergoes shear, where the deformation rate represents the speed gradient and the separation distance between edges. There is a proportionality relationship between shear rate ($\dot{\gamma}$) and external stress (τ) on the edges, and for this constant, the name of fluid viscosity (η) was given, as described by Eqs. 6.1 and 6.2

$$\dot{\gamma} = \frac{\mathrm{d}v}{\mathrm{d}x} \tag{6.1}$$

$$\eta = \frac{\tau}{\dot{\gamma}} \tag{6.2}$$

Fluids of technological interest that deviate from Newtonian behavior, that is, do not have a linear relationship between stress and shear rate, are called non-Newtonian fluids, and have viscosity values dependent on the shear speed. The most common cases of non-Newtonian fluids have reduced viscosity as the shear rate increases: These are called pseudoplastic fluids. When the fluid presents a stress value that determines the flow, it is a plastic fluid. This rheological behavior can happen due to several factors, such as the particles' physical characteristics and the additives that make up their suspension. In the case of suspensions, particles with a high surface area have a high tendency to agglomerate. These agglomerates are porous three-dimensional structures that absorb water within their pores and, with less liquid in the middle, there is a higher number of collisions between the particles, thus increasing the suspension's viscosity [5, 21]. When sheared, they undergo breaking mechanisms, releasing the water trapped inside the pores. In addition to the rupture of agglomerates, the phenomenon of pseudoplasticity can be seen in liquid systems whose molecules or polymeric chains are entangled at rest. With the application of shear, there is a preferential orientation of its chains to the flow. Finally, the application of shear breaks solvated molecules, which also trap water molecules in their structure. These three factors contribute to the appearance of pseudoplasticity.

Eventually, fluids may show an increase in the viscosity value as the tension/shear rate ratio increases. This behavior is called dilation. This behavior is characteristic of highly concentrated suspensions, where the particles are very well packed and very close together. To flow in this situation, the liquid must run through the channels between the particles. Although the flow of fluid between the particles is relatively easy under low shear rates, raising its intensity becomes increasingly difficult, as a result of the higher number of shocks between particles. In this way, all the factors that reduce the average separation distance between the particles and hinder the relative movement between them favor the suspension's dilating behavior.

Bingham's plastic fluids are those that have behavior like Newtonian fluids. They display a linear relationship of viscosity with the shear rate, but they also need a certain nominal tension to start their flow. The different rheological behaviors of fluids relating to stress and shear rate are shown in Fig. 6.4.

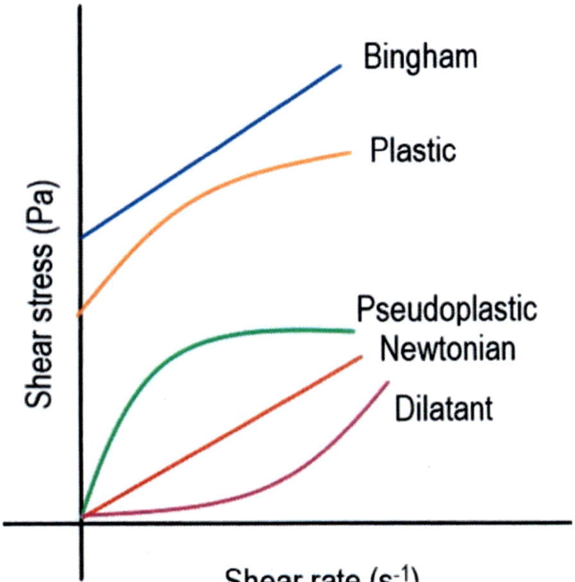

Fig. 6.4 Rheological classification of different types of fluids

The role of time in rheological characterization aims to understand flow mechanisms as the fluid is maintained at a given fixed shear rate. When viscosity decreases with the evolution of stress and shear rate, there is a pseudoplastic fluid. However, when effort ceases, it is called thixotropy. Thixotropic behavior originates when such suspensions are maintained at low shear rates and then are subjected to higher rates in a relatively short period. This is done to destroy the agglomerates present in the suspension. However, the breakdown of clusters requires a certain amount of time, and due to this fact, thixotropic behavior is observed in suspensions. The technological advantage of suspensions with thixotropic characteristics is that the suspension in its resting state has a higher viscosity, and when passing through stages of transport, pumping, or even its very use, there is a better flow condition. Different rheological parameters result in characteristic hysteresis curves. Consequently, shear history is an essential factor. Thixotropy is seen on the flow curve as the resulting area that comprises the entire rheological assay's rise and fall curves.

The opposite rheological behavior, in which viscosity increases with a constant shear rate, is called rheopexy, also known as anti-thixotropy. This behavior is not very common, but it is explained by processes of gelling and stiffening caused by redox transitions of surface species, the formation of hydrogen bonds or electrostatic unions. Another factor that can give rise to rheopexy is the form factor and the morphology of the particles (platelets or rods), and the use of dispersants with very extended polymeric chains that make up the suspensions [41]. Figure 6.5 shows typical curves for time-dependent behaviors.

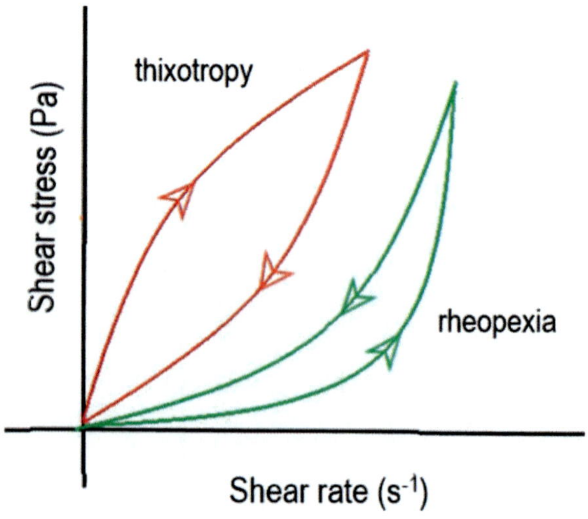

Fig. 6.5 Rheological behavior of time-dependent fluids

6.2.1.3 Granulation

In preparing the powders for the shaping step, especially when shaping will be carried out by compaction, a granulation step is necessary. The filling of the press matrices is not homogeneous, so reaching uniform densities in a reproducible way is impossible. As a solution to this problem, one may granulate the powders, that is, agglomerate the ferrite powders in a controlled manner to make them suitable for filling the mold.

The granules must have enough resistance to be manipulated but sufficient softness so that they can be compacted. Thus, there are several processes available in the ceramic industry that can be used for the granulation of ferrite powders: granulation by powder milling, incorporation of additives, stirring/evaporating a suspension, drying granulation from suspension, freeze-drying, or spray-drying [3].

The most efficient method of producing granules for compacting with controlled sizes and density is spray-drying. It consists of spraying a stable slurry in the form of drops in a stream of hot air (or inert gas) to produce practically spherical granules (50–500 μm). It is a continuous, economic, and reproducible drying process. If the suspension particles are well dispersed during spray-drying, the particles move toward the surface of the slurry droplet, forming a rigid shell with a high packing density. After the water in the shell evaporates, external pressure causes the shell to collapse inward, creating a spherical-hollow granule. The suspension control is decisive in obtaining the granules. There are several factors that affect the control of the product.

To increase spherical-solid granule production by controlling granule morphology, Mori et al. [22] investigated the relationship between slurry and spray-dried granule properties. They prepared various slurries using different MnZn ferrite particles and evaluated the products using flow curve and adsorbed dispersant amount measurements and sedimentation tests. Spray-dried granules were evaluated using morphology observations, density and strength measurements, and compression tests. The results suggested that the change in a slurry's particle dispersion and flocculation state during drying strongly influenced the granule morphology. Figure 6.6 shows the SEM images for each of the prepared slurries.

Yan and co-workers [23] showed a series of morphology-selected copper ferrite ($CuFe_2O_4$) anode materials for sodium-ion batteries prepared via spray-drying method using glycine as the complexant and accelerator. The excellent electrochemical properties can be attributed, among other factors, to a proper porous structure, thus improving the cycle life of the battery, as shown in Fig. 6.7, that illustrates the format of granules.

On the other hand, the lyophilization/freeze-drying method is a method that consists of a freezing and sublimation process of ice under a vacuum that transforms solid water directly into steam. The freeze-drying processing for granulation generally occurs in three stages: (i) freezing stage—the material must be cooled to ensure that the temperature is low enough for the solvent to sublimate in the operating pressure of the freeze dryer. The freezing speed influences the subsequent steps, as quick freezes result in tiny crystals, therefore, small and numerous pores. However, slower freezing speeds result in larger pores that facilitate the escape of water vapor

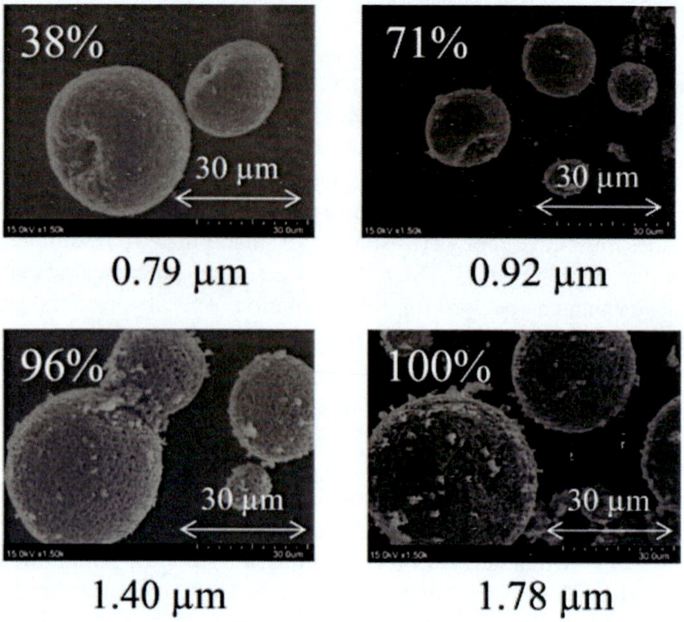

Fig. 6.6 Granule morphology observed by SEM (Reprinted from [22], with permission from Elsevier)

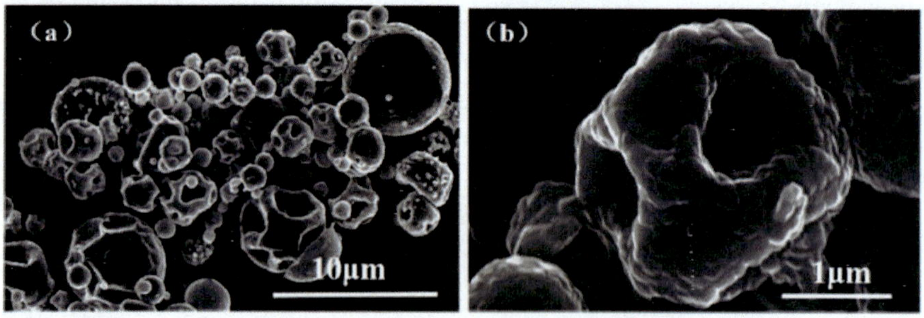

Fig. 6.7 FESEM images of granules obtained after spray-drying (Reprinted from [23], with permission from Elsevier)

in the drying steps, damaging the product structure [24]. There are several ways of freezing suspensions of ceramics, but the most common is the use of liquid nitrogen (−196 °C) or in freezer (−70 °C). A rotary evaporator, which rotates while the suspension is frozen, may be used to create spherical granules during the freezing stage. (ii) After freezing, the product is taken to the vacuum chamber of the freeze dryer. It is maintained under conditions of pressure and temperature of the solvent to be removed. Then, the primary drying will occur, which consists of the sublimation step up to 90% of the frozen solvent. (iii) The last stage of lyophilization is secondary

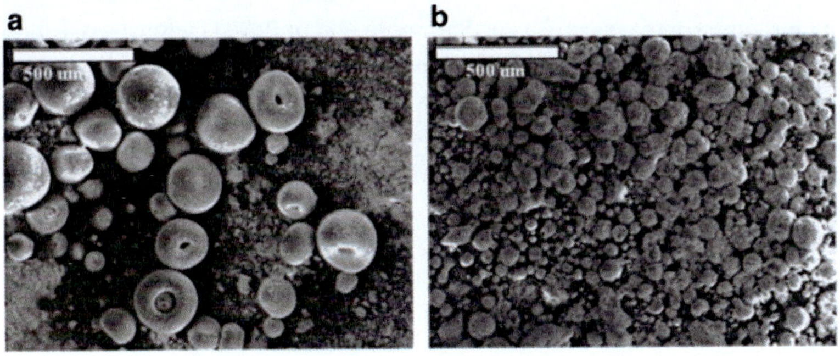

Fig. 6.8 SEM images of granules prepared by **A** spray-drying and **B** the freeze-granulation process (Reprinted from [25], with permission from Elsevier)

drying, which is responsible for removing residual moisture in the product even after the sublimation stage during primary drying. This remaining water is strongly adsorbed to the solid substrate and is also known as chemically bound water. In the end, granules of different shapes are obtained.

Researches [25] compared the spray-drying and freeze-drying processes for the granulation of MnZn power ferrites for subsequent compaction. In Fig. 6.8, the morphology of the granules can be observed for each of the two granulation processes. In the spray-drying process, the formation of large granules and the presence of intergranular voids are observed. In the freeze-drying process, small and similarly shaped granules are formed.

The different sizes and shapes of the granules can cause different compaction levels, which cause a change in the final properties of ferrites. In a way, modifying the granulation method can also be a form of tuning spinel ferrites. Spray-drying and freeze-drying are the most usual granulation techniques. Furthermore, they can be used in other agglomeration processes, not necessarily only for later compaction.

6.2.2 Shaping

The shaping methods are selected depending on the shape and complexity of the material to be manufactured, production volume, and final finish, among others. The forming process not only includes producing materials with desired shapes and dimensions but also producing compacts with higher density and uniformity. There are numerous shaping methods for ceramic materials, as well as ways to classify them. Spinel ferrites can be processed by several of them, depending on the needs required in the application.

In dry conformation, humidity is very low (<7% by mass). The powders can be compacted by applying pressure in a specific direction, normally axial and isostatic. In the plastic conformation, the humidity reaches between 15 and 50% (mass), which gives the material a certain plasticity, allowing the conformation through extrusion or injection molding (in this case, it is necessary to add polymers additives). Finally, the conformation from suspensions also is known as colloidal processing. In this case, the process starts from suspensions, usually colloidal, in which the liquid content can be higher than 50% [26]. The most common methods are based on filtration or casting mechanisms (slip casting, tape casting), flocculation/coagulation, gelation, and deposition.

6.2.2.1 Uniaxial and Isostatic Pressing

Uniaxial pressing is widely used in large industrial productions of spinel ferrites. It consists of applying a uniaxial pressure in metallic molds in which the already granulated powders containing low humidity are accommodated. The material is pressed against the mold cavity, pressed by a plunger in a uniaxial direction, and then extracted by a lower plunger.

One of the most critical stages of pressing, as already discussed, is filling the mold. For it to be filled correctly, the powders must show good fluidity. Besides, the smaller the particle size, the more contact the grains have with each other and the worse their ability to flow. For this reason, granular powders should always be used, as they have better flowability and are easily broken during compaction, when pressures between 20 and 100 MPa are applied. In the process, while pressure is acting on the system, there is an initial rearrangement of the granules. Then, deformation and/or fragmentation begins, when the elimination of macroporosity occurs. Finally, the elimination of microporosity present inside the granules occurs.

Some additives can be used in the pressing process, as moisture is hardly sufficient to shape the bulk properly. Spinel ferrite powders, in general, have little plasticity. Polyvinyl alcohol, polyethylene glycol, and some polysaccharides can be used as aids in the agglomeration process. As lubricants, magnesium stearate, talc, oleic acid, or zinc stearate may be also used.

Even with the utilization of additives, care with obtaining powders with good fluidity, and uniform filling of the mold, the final products may often display a series of defects. These defects associated with the pressing process include differences in density along the sample, leading to the appearance of voids and cracks, and finally, delamination, unevenness of edges, or rupture of edges. Isostatic pressing may be used to improve this condition.

In isostatic pressing, the application of pressure occurs in a uniform hydrostatic form in a powder contained in a flexible polymeric mold (natural, butyl, nitrile, or polysulfide rubber, neoprene, silicones, polyurethane, flexible polyvinyl chloride). Pressure can vary between 35 and 1380 MPa in laboratory conditions and <400 MPa in industrial conditions. The pressure applied to the fluid is transmitted uniformly in all directions. The great advantage of this process concerning uniaxial pressing is a significant reduction in friction between the granules and the mold walls. The pieces obtained by this method have better packing, less density gradients, and more uniform products.

The isostatic pressing technique can also be carried out hot. The technique is known as hot isostatic pressing (HIP). Some equipment can reach up to 1800 °C.

6.2.2.2 Slip Casting

The slip-casting process fills a porous ceramic mold (in most cases, made of plaster) with a concentrated suspension with adequate fluidity. During the drainage of the liquid, a layer is formed on the mold walls because of capillary pressure. The drainage of the liquid can be aided by positive or negative pressure (vacuum) that helps increase the material filling rates. After the solvent's total suction, the deposited particles form a moist green body that is subsequently dried.

This green body is then sintered. To allow handling the dry green body and reduce cracks during drying, a polymeric additive, the binder, is usually added to the suspension. The binder adds plasticity to the wet green body and gives strength to the dry green body. Considering the case of spherical particles, the mechanical strength of the green body depends on the concentration of the binder: the binder must be fired without creating cracks or other deformations. Thanks to the binder's sacrificial nature, it can create dense ceramics or create well-defined internal pores or channels [5, 27].

The advantage of slip casting is that, from a deflocculated suspension, it produces complex shapes at a low cost. On the other hand, the formation of agglomerates or sediments during the process can lead to local anisotropic shrinkage leading to cracking, warping, and deformation in the drying and sintering phases.

Thus, the desirable properties for the suspensions to be cast are good fluidity for the entire filling of the porous mold; low sedimentation speed when it is at rest; uniform drainage condition; quick mold separation; at samples must have good green strength; low retraction during drying; and being bubble-free.

The parameters to be controlled in the process are the raw materials, the rheology of the suspension while filling in the mold, the leakage rate, and the suspension's density and consistency. Figure 6.9 illustrates the slip-casting process.

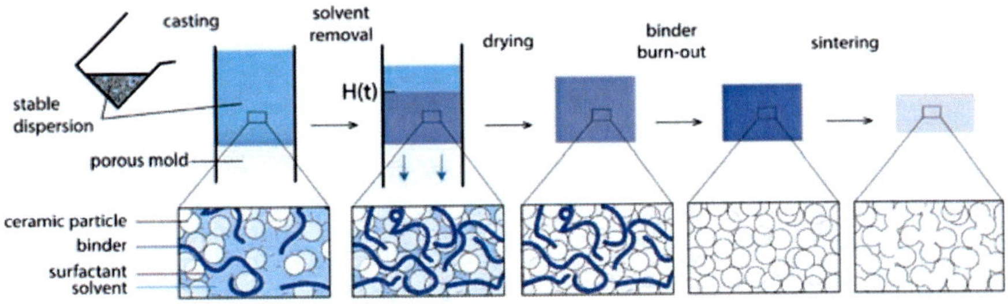

Fig. 6.9 Schematic representation of slip-casting processing (Reprinted from [28], with permission from Elsevier)

In the slip-casting processing, an external magnetic field may be applied. This method offers the advantage of orienting the preferential crystalline axis of the spinel ferrite grains. Besides, it is possible to use this methodology to create textures and orient diamagnetic ceramic powders. Most of the ceramic phases are diamagnetic. The application of an external magnetic field to a diamagnetic material induces an internal magnetic field in the other direction, causing a repulsive force.

With these oriented structures, many properties of the materials may be improved. Anisotropic structures are known to lead to anisotropic properties that significantly improve mechanical properties, for example. Strength and toughness can be increased along the direction of reinforcement, while toughness is increased along the perpendicular direction [28].

6.2.2.3 Freeze Casting

As already described in Sect. 6.2.1.3, granulation, the "freeze casting" technique comprises preparing a ceramic suspension (aqueous or non-aqueous) and freezing this suspension, followed by sublimation of the solvent under vacuum. In the case of obtaining ceramic pieces, a molding step is added before freezing. Finally, sintering is carried out, which consolidates the ceramic walls while maintaining the pore structure, thus preserving a replica of the structure frozen by the solvent. The four basic steps of freeze casting are illustrated in Fig. 6.10.

Freeze casting has attracted much attention in the production of porous materials, as it is a technique of low cost and little impact on the environment, as the solvent

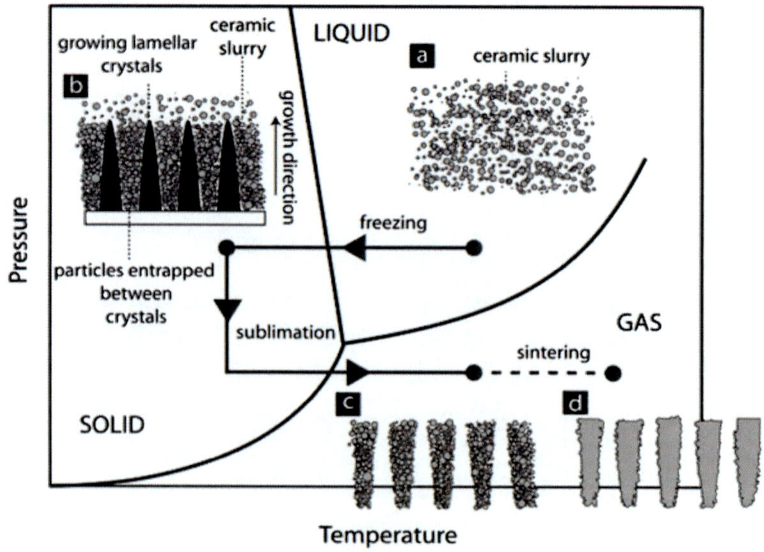

Fig. 6.10 Freeze casting process (Reprinted from [28, 29], with permission from Wiley)

temporarily acts as a binder, maintaining the desired structure after removal from the mold. The low addition of binders means that shorter times are used evaporation of additives, thus reducing the time at high temperatures. The most significant benefit of the technique is the possibility of producing porous ceramics with greater mechanical resistance than those obtained through other routes. It is worth mentioning that countless materials have already been prepared by this principle, demonstrating that this technique is not dependent on the type of material, as it is more related to physical interactions than to chemical ones. The four basic steps of freeze casting are illustrated in Fig. 6.11.

Magnetic filters can be produced using the freeze casting technique. Aguilera and collaborators [30] produced magnetic filters made of zinc ferrite nanoparticles,

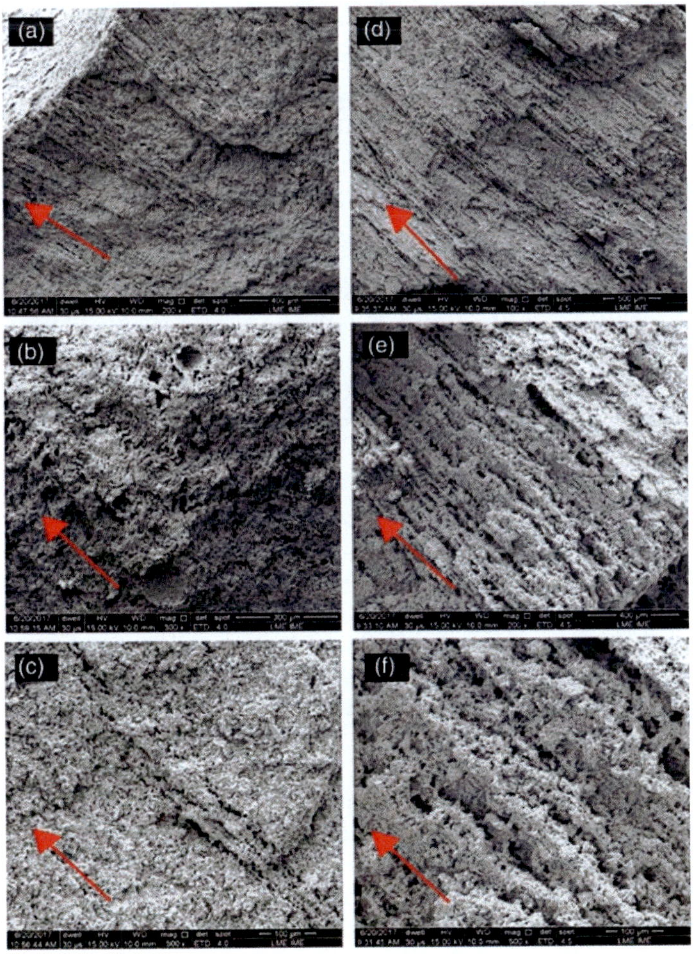

Fig. 6.11 Filters made of zinc ferrite nanoparticles, using the freeze casting technique (Reprinted from [30], with permission from Elsevier)

using the freeze casting technique with up to 3.0% by weight of PEG—8000 as a binder. The sintered porous bodies showed porosities of 48.0% and 63.0%. The authors confirmed the formation of porous channels in the direction of freezing. The formation of lamellar structures, characteristic of the growth of ice crystals, is also evident, with a columnar freezing front essential for application in filtration systems. The presence of homogeneous spherical pores and reticulated pores with freezing front alignments are shown in Fig. 6.11 (b).

6.2.2.4 Gelcasting

The gelcasting process consists of introducing organic monomers to an aqueous suspension of the ceramic powder, which polymerizes in situ, producing a three-dimensional lattice that consolidates the ceramic matrix. After forming in a mold with the final geometry of the part (near net shaping), a gelled structure is formed, which provides rigidity to the bodies in the green state.

To produce cellular materials, this process was associated with the aeration of the ceramic suspension containing bubbles. After the system has gelled, the ceramic is then dried and subjected to heat treatment, where densification and sintering occur.

In the gelation stage, when the polymerization of the organic monomers occurs, the high reactivity of free radicals causes the reaction with oxygen molecules present in the medium, reducing the kinetics of initiation and propagation [31]. The period between the addition of the reagents and the initiation of the polymerization reaction is called the induction period or inactive time. This period, controlled by changing the initiator and catalyst concentration, represents the time available to pour the aerated suspension into the mold before the polymerization and stabilization of the cell structure. In the reaction, a three-dimensional network is formed with branches and cross-links constituting the microgels that come together to form a macrogel network. The mesh formed inside the suspension keeps the particles trapped. The reaction occurs over the entire length of the suspension, providing gels with excellent homogeneities [32]. The process results in porous and rigid structures with complex shapes without intensive machining steps [33, 34].

The suspension must present a high content of solids and low viscosity to decrease the shrinkage during drying and present a good microstructure in green, forming low-density foams. The dispersing agent is added to promote the homogeneous distribution of powder in the suspension. Water is the most used solvent in this process. The most used monomers of low toxicity (compared to the acrylamide system) are methacrylamide and hydroxymethylacrylamide. As cross-linkers, methylenebisacrylamide and polyethylene glycol dimethacrylate stand out. Ammonium persulfate and tetramethylethylenediamine are, respectively, the initiator and the catalyst often added to trigger free radical polymerization. The relationship between monomers and cross-linkers varies between 2:1 and 6:1, influencing the gelation speed and the rigidity of the sample gel.

Ceramic parts of different structures and sizes have been successfully made by gelcasting using a wide range of materials. The use of gelcasting for making soft

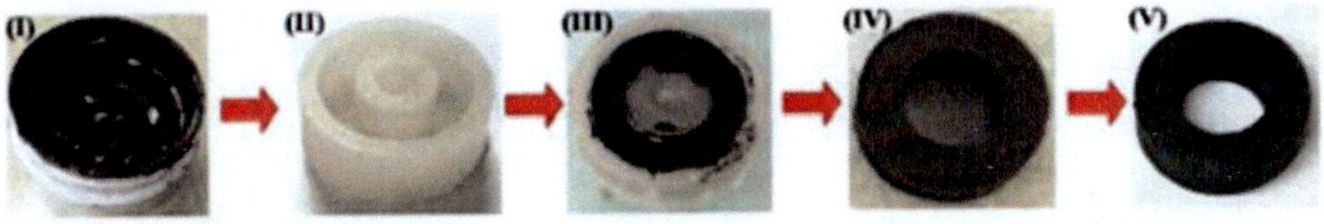

Fig. 6.12 Illustration of a process flow for making a toroid core by gelcasting: **I** LSF 50 slurry; **II** wax mold; **III** slurry injected into the wax mold; **IV** cured green body after demolding; and **V** toroid core after sintering (Reprinted from [35], with permission from Elsevier)

magnetic materials such as NiZn and NiCuZn ferrites has been increasingly stimulated. The applications involve components for micro-electro-chemical systems, for example. Liu and collaborators [35] used the pressureless gelcasting process to fabricate a NiCuZn ferrite to simplify the integration of magnetic components into power electronics converters. A commercial NiCuZn powder was gelcast into toroid-shaped cores and then sintered; the process is shown in Fig. 6.12.

Because of its ease, flexibility, and scalability, gelcasting has the potential to enable the integration of novel designs of magnetics for shaping and coupling magnetic fluxes aimed at achieving higher efficiency and power density of high-frequency power electronics converters.

6.2.2.5 Additive Manufacturing

Additive manufacturing or 3D printing is the generic term given to all technologies that produce parts automatically through the progressive addition of volume elements, which are usually layers of the same thickness. They are technologies that integrate two sub-processes: the physical production of each layer and the union of these layers in sequence to make the final piece.

Some of the essential characteristics of additive manufacturing are: The geometry of each layer is obtained directly from the data of the computer-aided design; there are no specific tools for each part geometry, and there is no need for tool changes, which reduces costs and speeds up the production of small quantities of parts; the properties of the material used for manufacturing are defined during the process; the pieces can be obtained in different orientations; and designers can modify and adapt their prototypes more quickly, resulting in faster and more effective product development cycles.

There are several commercial methods of additive manufacturing, with varying characteristics, capacities, and quality of the final product. Additive manufacturing can rapidly prototype innovative designs of magnetic components used in electronic power converters.

To print a magnetic multi-permeability core, additive manufacturing is necessary to extrude different magnetic materials in different locations of the magnetic core. In this sense, the magnetic raw material must be compatible with an extrusion-type 3D printer, have sufficient functionality to make functional magnetic cores, and have

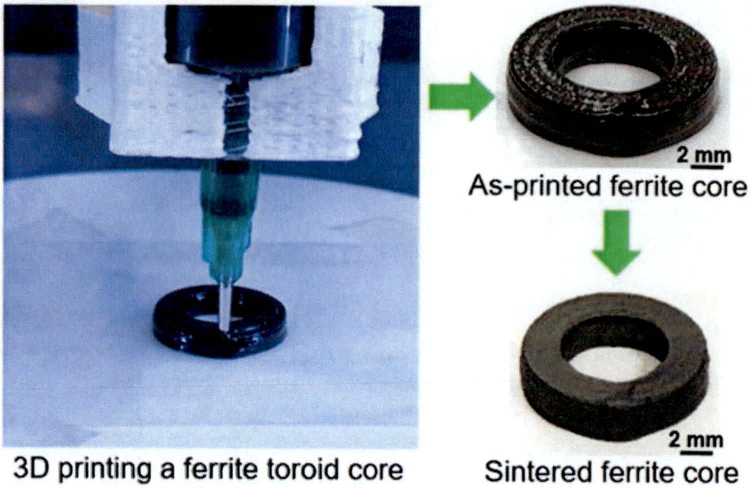

Fig. 6.13 Demonstration of a 3D-printed toroid core by the ferrite feedstock (Adapted from [36], with permission from Elsevier)

different permeabilities. Ideally, the raw material's permeability can be easily adapted to different values according to different needs. Ferrite can be produced from additive manufacturing with a prediction of magnetic permeability. Different amounts of silica can be used in the printing paste, together with the necessary binders [36]. Figure 6.13 shows the 3D manufacturing system, as well as some printed materials.

6.3 Films

The use of surface coatings to modify specific properties of certain materials has been intensifying in recent decades. The purpose of applying these coatings in the form of films may be the need for increased hardness, greater protection against corrosive attacks, and greater resistance to mechanical wear, among others. From the 1980s onwards, the study of thin ceramic films took off due to the growing need for the miniaturization of high-tech devices. Among the advantages presented by these materials are the smaller size, lower weight, easy integration with the integrated circuit technology, low operating voltage, high response speed, and the possibility of fabricating structures at the microscopic level [37].

Films can be classified into nanometric, micrometric, and millimetric, referring to the ranges of 0.1–100 nm, 0.1–100 μm, and 0.1–100 mm, respectively [38, 39].

With the development of high-frequency and magnetic and magneto-optical memory devices of nanoscopic dimensions, there has been a significant increase in interest in ferrite thin films in recent years. There are many applications of ferrite thin films for which low density and nanosized materials are required. Properties of ferrite thin films are often different from those of the bulk for various reasons, such as crystallite size, high defect density, grain boundaries, texture, constraints caused by the substrate, and fabrication parameters.

Ferrite thin films have attracted much attention in recent years as one of the candidates for high-density magnetic recording and magneto-optical recording media because of their unique physical properties such as high Curie temperature, magnetic anisotropy, magnetization, and excellent chemical stability.

The sol–gel synthesis is one of the most used methods for preparing thin films due to the possibility of controlling the thin film microstructure. Tape-casting, spin-coating, dip-coating, electrophoretic deposition, and spraying methods are also widely applied. Further, technologies such as electrophoretic deposition are becoming increasingly important in recent years.

Films with thicknesses exceeding 1 μm are known as thick film, whereas thicknesses below 1 μm (or even below 100 nm) are known as thin films. When a spinel-type ferrite powder is deposited with a thickness of more than 1 μm, the product is known as spinel-type ferrite thick film. These materials are mainly used in optical and electronic applications such as surface mount devices, hybrid integrated circuits, and sensor circuits.

6.4 Tape Casting

Ferrite films have practical applications in the field of magnetic recording, magnetic sensors, and multiferroic devices. Due to the demand for integrating ferrite materials into a single semiconductor package for manufacturing smaller electronic devices, the manufacture of thin or thick ferrite films has been in evidence in recent years.

Tape casting is a technology used to produce thick ceramic films with a thickness of 25 μm–1 mm. It is a technically simple and economically viable method when compared to other methods of film production. The process consists of preparing a suspension of inorganic powders in an aqueous or non-aqueous liquid medium. This medium is composed of many additives: solvents, plasticizers, deflocculators, and binders. The suspension is cast on a flat sliding surface, with a passing blade leveling the suspension, resulting in a controlled film width and thickness. When the solvent is evaporated, the particles coalesce to give rise to a green body with a relatively high green density. This piece is flexible and can be rolled, cut, punched, or laminated. Subsequently, the material is thermally treated to dispose of the organic substances and fired [20, 40, 41]. Figure 6.14 shows the schematic illustration of a tape caster.

In the tape-casting process, particle aggregation due to van der Walls attraction damage slurry stability and cause viscosity increases. As for magnetic particle slurries, an additional magnetic attraction between particles exists. However, well-dispersed magnetic particles are usually produced via a surface modification using the colloidal processing concepts to disperse particles.

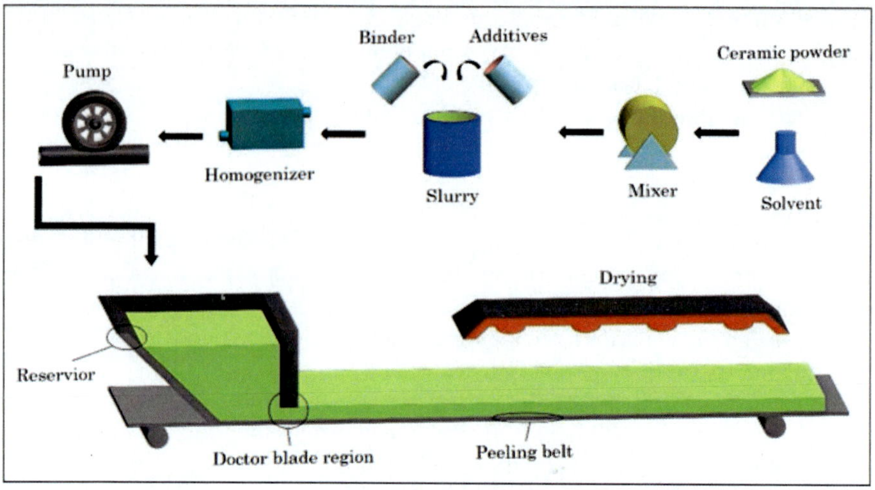

Fig. 6.14 Scheme of process and equipment (tape caster) for continuous processing of bonding ceramic tapes (Reprinted from [42], with permission from Elsevier)

Afterward, the binders, plasticizer, and dispersant are added to the ferrite paste, and these compounds are mixed by ball milling. These prepared compounds are used for fabricating ferrites with a thickness of 10 μm by the tape-casting method.

The great aim of researchers who work with spinel ferrites in the processing of tapes by tape casting is manufacturing materials with functional grading. These materials show a gradual variation in their properties from one end to the other of the ceramic. The technique of tape casting is selected, in this case, due to the production of large-area, fine, and flat ceramics, which can be standardized, stacked, and laminated to form three-dimensional structures. Multiple pastes can be shaped into tape to form a single ribbon of functionally graded ceramics. These functionally graded ceramics are used in magnetic refrigeration processes in which there is a temperature gradient across the piece [42] (Fig. 6.15).

6.4.1 Electrophoretic Deposition (EPD)

Among all routes to obtain thin films, electrophoretic deposition stands out as a powerful and versatile colloidal process to address the need for economic and mass production of films using suspensions with a low concentration (<1 g/l). This drastically increases its range of applicability, competing with other non-colloidal techniques. The electrophoretic deposition was discovered in 1808 by the Russian scientist Ruess. It was first used in a practical application in 1933 to deposit thoria particles onto a platinum cathode [43].

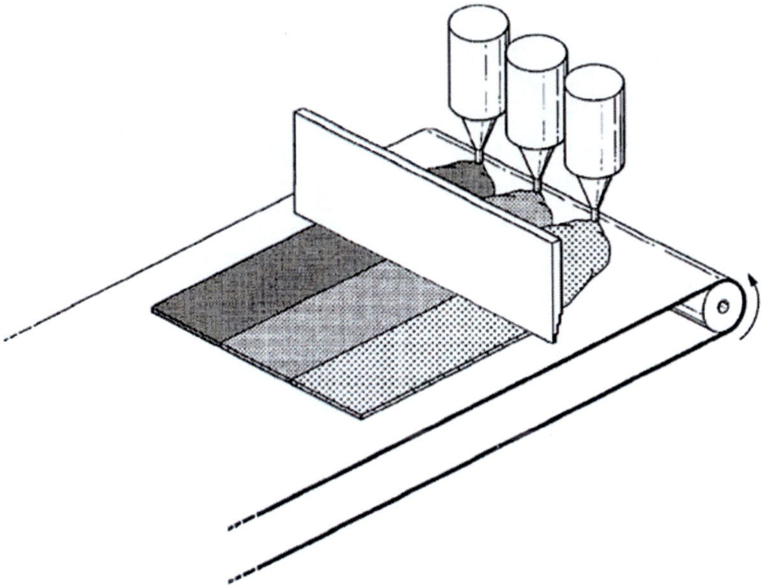

Fig. 6.15 Schematic illustration of the newly developed method of tape casting to produce functionally graded ceramics for magnetic refrigeration applications (Reprinted from [42], with permission from Elsevier)

Electrophoretic deposition is a term for a broad range of industrial processes, including electrocoating, e-coating, cathodic electrodeposition, anodic electrodeposition, and electrophoretic coating.

In most of these coating processes, the transport of small particles is mainly governed by Brownian motion, resulting in entirely random particle–particle and particle–substrate collision. A directional force (external electric field) is applied in electrically driven processes, leading to 2D and 3D ordered structures at the nanoscale. In other words, the electrophoretic deposition utilizes the mobility of charged particles in solution upon exposure to an applied electric field to deposit those particles onto an electrode that acts as a substrate. This technique offers high deposition rates using a simple apparatus that is easily scalable for production. Processing parameters of importance for electrophoretic deposition are particle mobility (zeta potential, solid content, viscosity), solvent, applied electric field, and deposition time. After solvent evaporation, a cohesive layer of particles is formed, reproducing the shape of the working electrode. Consequently, deposit yield and particle packing in electrophoresis depend on the stability and solids content of the suspension and the aggregation and arrangement of the particles on the electrode surface. The performance of electrophoretic deposition in producing homogeneous and reliable films strongly depends on the particle surface chemistry, the behavior of the surface–liquid interfaces under an electric field, and the development of particle–particle and particle–substrate networks during particle assembly [44]. Figure 6.16 shows the

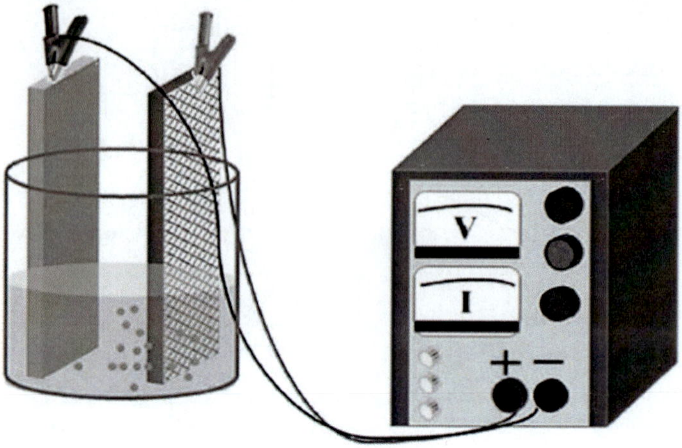

Fig. 6.16 Schematic diagram of an EPD equipment setup

schematic illustration of an electrophoretic deposition system.

The parameters of the electrophoretic deposition process are fundamental in the final properties of the formed films. Several factors can influence the permanence of the ferrite spinel film on the substrate. Several researchers have been trying recently to find the best conditions for deposition in each ferrite system in this context. Investigations are being conducted to assess how the voltage and the separation distance between the electrodes influence the thickness and the homogeneity of the deposits prepared from suspensions with different zeta potentials and different concentrations of particles with various sizes. The thickness of the deposits increases with the increasing zeta potential of the suspension and the growing separation distance between the electrodes.

It is worth mentioning that reaching ferromagnetic and ferrimagnetic films via electrophoretic deposition is challenging since a strong magnetic attraction is present in these suspensions even in the absence of an external magnetic field. Therefore, the necessary precautions with colloidal processing must be taken.

Hashi and collaborators [45, 46] fabricated Ni–Zn ferrite films by the EPD method on an ITO glass substrate from an aqueous solution at room temperature. They evaluated the particle's surface potential and verified that the ferrite particles (10–30 nm) developed a considerable positive surface potential. The thickness of the film was 12 μm fabricated at 2.7 and 1.2 V for 30 min, giving a deposition rate of approximately 400 nm/min.

6.4.2 *Spin-Coating*

Spin-coating is a rotation coating technique widely used to deposit thin films on flat substrates, creating films with uniform thickness at the nanoscale. This technique consists of adding a predefined amount of a precursor suspension over the center of a substrate and subjecting the system to a centripetal force to spread this suspension, forming a thin coating on the substrate [47]. The centrifugation process can be divided into four main stages:

a. deposition: addition of the precursor suspension on the substrate;
b. centrifugation: the substrate rotates on its axis, and the liquid flows radially toward the edges due to the centripetal force;
c. drainage: the excess liquid is eliminated through the edges of the substrate in the form of drops; and
d. evaporation: the remaining liquid evaporates, and the dry coating is fixed on the surface of the substrate, as shown in Fig. 6.17.

The spin-coating deposition technique has important characteristics in the production of thin films, such as high deposition rate, thickness homogeneity, stoichiometry over large areas, nanostructuring possibility, and low cost. The technique is simple and depends on low-cost equipment (spin coater). It is method of coating by rotation widely used to deposit thin films on flat substrates. Moreover, films in the 100 nm range (which are not easy to prepare by other methods) are easily obtained by this technique.

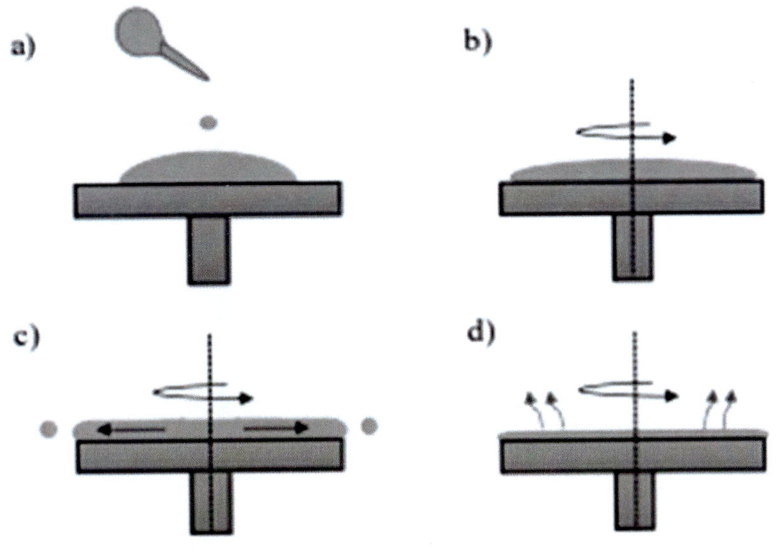

Fig. 6.17 Scheme of the main steps of the spin-coating technique. **a** deposition; **b** centrifugation; **c** drainage; **d** evaporation

The speed at which the system rotates influences the precursor layer on the substrate; the higher the angular speed, the thinner the film layer will be. The thickness of the film also depends on the viscosity of the precursor suspension.

There is a growing interest in ferrite thin films with controlled functionalities for advanced applications. $Ni_{1-x}Co_xFe_2O_4$ thin films, for example, may exhibit magnetic, sensing, and catalytic properties as a function of Ni and Co concentration in the spinel structure. Using a spin-coating method, Tudorache and Co-workers [48] obtained a 200 nm thin film spinel structure of nickel–cobalt ferrite. Thin films obtained by spin coating, due to their characteristics, can be used as gas and catalysis sensors. In other work [49], $CoFe_2O_4$ nanoparticles were deposited on porous Si using a spin-coating method. The combination of ferrite nanoparticles and porous silicon opens up the possibility to obtain multifunctional nanocomposites with various properties. Thus, when applied on silicon, they are favorable candidates for several potential applications, such as anti-reflective coatings and high-reflectivity mirrors, optoelectronics, catalysis, gas sensors, electronics, and biomedicine. Moreover, the spin-coating method can be used to form multilayer composites. The facile spin-coating techniques allows the preparation of bilayer films with smooth surfaces even when many individual coating steps are applied. In this way, Quandt and co-workers [50] obtained smooth and uniform bilayer composite films of $BaTiO_3$ and $CoFe_2O_4$ in $SrTiO_3$ single crystals by spin coating (Fig. 6.18). They achieved ten and five individual coating steps which resulted in a thickness of about 250 nm for

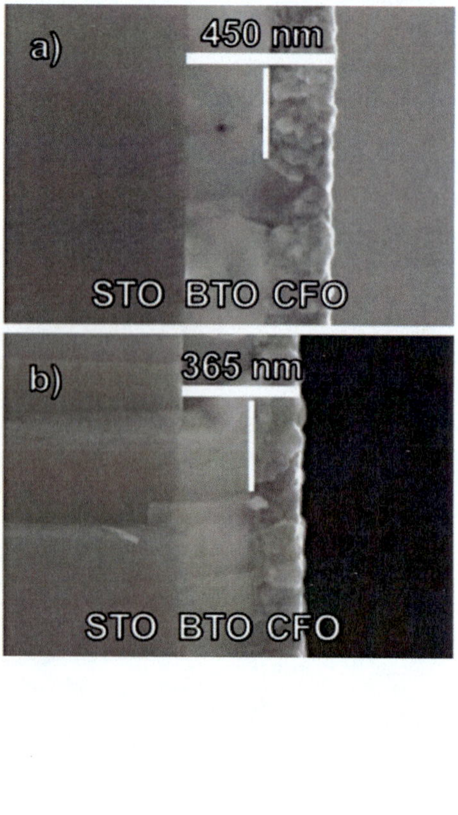

Fig. 6.18 Cross-section SEM images of the BTO/CFO ($BaTiO_3$ and $CoFe_2O_4$) film at different positions of the STO ($SrTiO_3$) substrate **a** edge and **b** center (Reprinted from [50], with permission from Elsevier)

$BaTiO_3$ and 200 nm for $CoFe_2O_4$, respectively, and a total film thickness of about 450 nm. The layers of both components are highly dense.

6.4.3 Dip Coating

Dip coating has often been used to produce thin films on different substrates. Figure 6.19 shows the main steps involved in the process that can be divided into:

a. immersion of the substrate in the suspension;
b. emergence: extraction of the substrate at a constant speed; and
c. deposition, drainage, evaporation, and drying of the covering.

The principle of this technique consists of immersing a substrate inside the "precursor" suspension and then removing it from the liquid. The substrate emersion stage, after the immersion stage, drags part of the suspension, generating a mechanically fixed fluid covering. In this process, the speed must be controlled, constant, and predefined, to guarantee that the forces resulting from the system are balanced, allowing a homogeneous deposition on the substrate. The interactions between the substrate and the suspension and the immersed substrate residence time will also influence the final characteristics of the film. The thickness of the coatings can be determined by the forces present during the deposition step. The main effects can be described as: dragging force parallel to the emergence speed (proportional to viscosity and emergence speed); gravity force acting in the opposite direction of the drag force; resultant strength of surface tension; inertial force of the boundary layer of the covering to the substrate (same direction as the drag force); surface tension

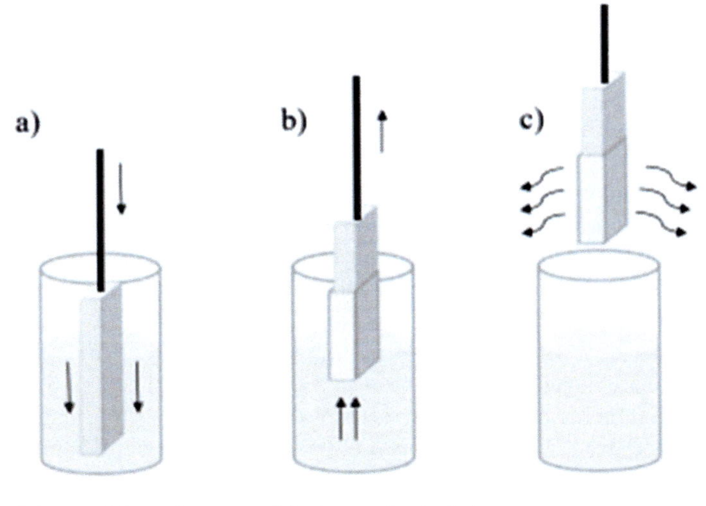

Fig. 6.19 Scheme of the main stages of the immersion technique

gradient (generated by additional forces); interactions between the substrate and the suspension [51]. At the end of the evaporation and drying process, a solid coating is achieved. In this step, the most crucial factor is the diffusion of the vapor from the surface of the coating. To increase the thickness of the film or to deposit a different material on the existing film, the immersion process may be repeated several times. This technique also allows the deposition of film layers on both sides of the substrate, besides being relatively easy and presenting a low cost.

As is already known by this point in the book, the structure–property correlation of thin films of spinel ferrites depends on their composition, preparation conditions, particle size, and morphology. The properties, including the magnetic ones, of thin ferrite films are also strongly dependent on their nanostructure and deposition methods.

High-quality homogeneous nanocrystalline thin films of $CoFe_2O_4$ were manufactured using dip coating. The films annealed above 500 °C exhibited a crystalline phase with an inverse spinel structure. The prepared films were ferrimagnetic and increased their saturation magnetization with the annealing temperature [52].

6.4.4 Spray Pyrolysis

The technique of deposition of thin films by spray pyrolysis utilizes a spray of an aqueous solution containing soluble salts (cations) of the atoms of the desired compound onto a previously heated substrate. The volatile reagents will vaporize and thermally decompose on the substrate surface, with nucleation and growth of the film. This technique is considered quite simple with a low-energy cost, and it is commonly used to cover large areas.

During the formation of the film, several steps occur sequentially or simultaneously, and the process can be divided into three phases: atomization of the precursor solution, aerosol transport, and decomposition of the precursor solution on the substrate.

The structure of the film to be deposited and its morphology, whether dense or porous, are strongly influenced by the deposition parameters. If intermediate reactions of deposition via spray pyrolysis occur at a certain distance from the substrate, this film will be porous and with some cracks. If the same reactions occur closer to the substrate, without touching it, the film will be dense. But, if these intermediate reactions occur long before touching the substrate, a powdery film will be produced.

In addition, the technique can be classified according to the type of atomizer and the operating energy. Moreover, it can be based on the energy of the gases.

Although there is guided gas transport in the tube and expanding hot air over the substrate, a significant portion of the aerosol droplets is lost in turbulence, which tends to reduce the low-cost advantage especially for the use of more expensive precursors.

This technique is not limited to generating only nanocrystalline materials; it also allows us to control the morphology of the particles through the process parameters.

Its main advantages are high purity, better uniformity, greater stoichiometric control, homogeneity in chemical composition, ability to produce small spherical particles with low size distribution, and ability to synthesize a variety of multicomponent materials. The use of this technique for the manufacture of films can be applied to a larger surface area and in non-flat structures under ambient conditions, an impracticable feature for techniques such as dip or spin coating.

Spray pyrolysis is a versatile and unique technique that produces extended thin films with high surface area. The properties of the produced thin films depend strongly on the preparation conditions, such as molarity of a solution, deposition temperature, spray rate, distance between the nozzle and the substrate, annealing temperature. It is worth mentioning that the deposition temperature and the annealing temperature are the most critical parameters, as they have a strong relation with the film's roughness, cracking, and crystallinity.

One of the advantages of the spray pyrolysis method based on inorganic metal precursors is that it allows the preparation of ferrites at temperatures below 650 °C [53]. Some disadvantages of films made with this technique are thickness is not uniform and the size of the atoms forming the film is limited and the film in chemical solution to grow must be homogeneous. A schematic drawing of the spray pyrolysis technique is shown in Fig. 6.20.

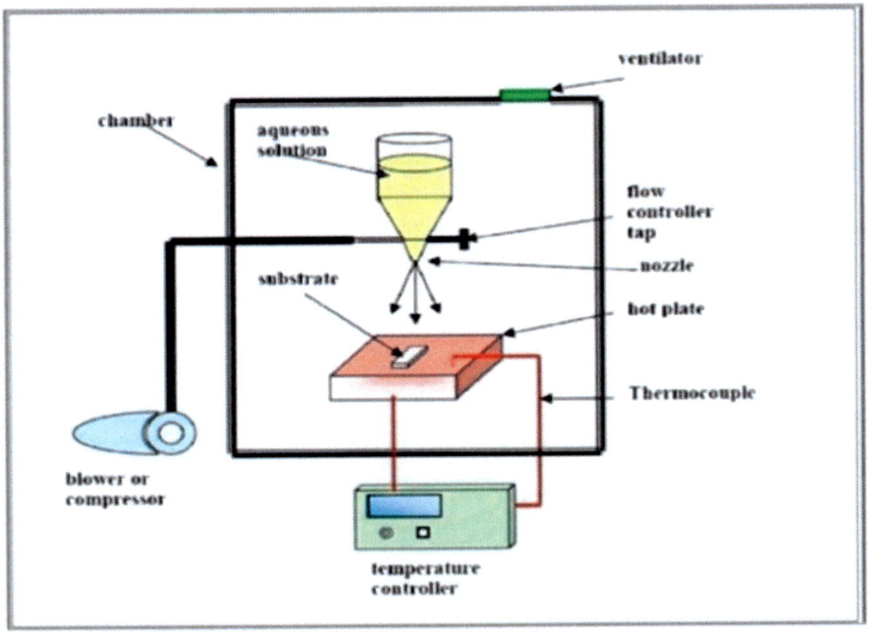

Fig. 6.20 Schematic drawing of the spray pyrolysis technique (Reprinted from [54], with permission from Elsevier)

Several ferrite compositions have already been used to obtain thin films, demonstrating the efficiency of the technique for a number of applications [22, 25, 53, 55–59]. Spray pyrolysis deposition is a versatile technique for producing various thin film materials of a wide range of the composition in air at a relatively low temperature and a high deposition rate.

Gas sensors (SO_2) of $Cu_xZn_{1-x}Fe_2O_4$ composition were obtained from spray pyrolysis deposition [60]. The variation of the synthesis parameters, mainly in the substitution of Cu, resulted in a change in the operating temperature of the film sensors. The thin film sensor of pure zinc ferrite shows the ideal operating temperature of 150 °C, while the proper replacement of copper leads to a reduction in operating temperature to 120 °C with a higher response. Sensors that operate at a lower temperature are attractive for detecting some gases such as SO_2, as they are more efficient and have low-energy consumption.

6.4.5 CVD (Chemical Vapor Deposition) and PVD (Physical Vapor Deposition)

In gaseous processes, CVD (Chemical Vapor Deposition) and PVD (Physical Vapor Deposition) methods are used to obtain thin films. In the chemical vapor deposition process, the film is formed by means of chemical reactions between reactant gases (precursors). The substrate is exposed to one or more precursor gases, which are injected into a camera. Unwanted by-products are also formed, which are removed by means of a gas flow during the process. Large areas can be coated, and the process and the thickness of the film can be controlled by the deposition time [37].

Very homogeneous, smooth, and dense films with low roughness can be obtained through chemical vapor deposition. Figure 6.21 shows a sample obtained using this technique. The SEM image exhibits a clear and sharp interface between the $CoFe_2O_4$ film and the $MgAl_2O_4$ substrate. The film looks very smooth and dense, without any cracks in the entire cross section or surface of the film [61].

The physical vapor deposition process is a deposition technique in which the material to be deposited is vaporized from a solid or liquid source, in the form of atoms or molecules, and transported in the form of a vapor through a vacuum or plasma to the substrate where it condenses. By this process, films with nanometric thicknesses can be obtained. Most common among the main techniques for obtaining PVD coatings are thermal evaporation (heating a material until it forms a vapor that condenses on a substrate and thus forming a coat) and sputtering (electrical generation of a plasma of the species of the coating) methods.

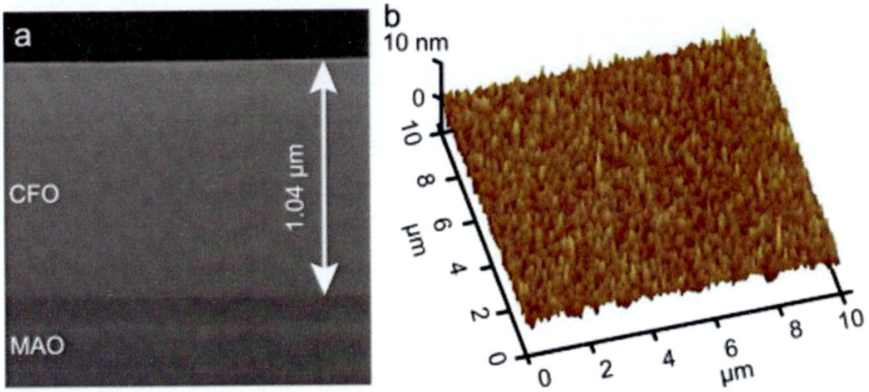

Fig. 6.21 **a** SEM image across and **b** AFM image of the surface of a $CoFe_2O_4$ film grown on $MgAl_2O_4$. Both images show that the film has a very low surface roughness (Reprinted from [61], with permission from Elsevier)

6.4.6 *Sputtering*

The sputtering technique basically consists of a vacuum chamber where two electrodes form a plasma in inert gas transferring the target material (target), known as a cathode, to the substrate fixed on another electrode (anode). Such technique allows good control of chemical composition as well as good reproducibility, which justifies its use in the semiconductor industry. Spinel ferrite films produced by this technique have promising properties for applications in RAM memories and applications in memory devices. The reactive magnetron sputter technique allows a wide variation in the composition of the film and perfect adjustments of composition and stoichiometry of spinel ferrites. Co-spraying of two different metal targets allows adjustment of the metal composition, while the oxygen content of the layers can be adjusted by the controlled addition of oxygen to the sputtering gas. In addition, by modifying the substrate heating parameters, for example, it is possible to achieve crystalline phases at temperatures below 600 °C [62]. Figure 6.22 shows an example of a system for production of batch films on an industrial scale. The device is equipped with four magnetron sources, a rotating substrate holder with sub-rotation, a hollow cathode plasma source, and a central heater.

6.4.7 *Pulse Laser Deposition (PLD)*

Pulse laser deposition consists of a mechanism similar to sputtering. However, plasma is formed through the laser that strikes the target, tearing off the ions due to incident energy from the laser on the target. This technique has the advantage of depositing films with the same chemical composition as the target.

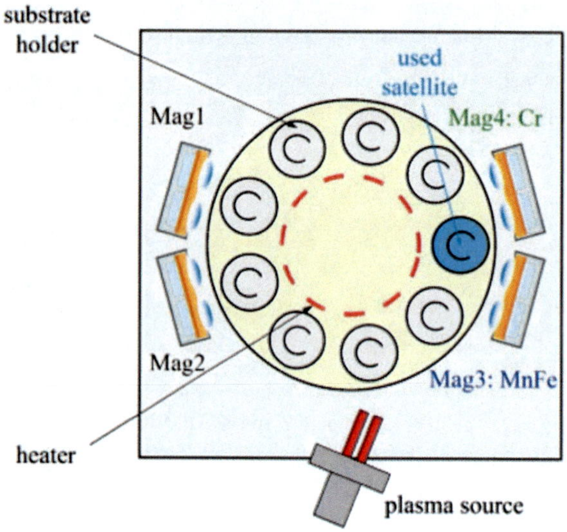

Fig. 6.22 Schematic view of the deposition sputtering arrangement (Reprinted from [62], with permission from Elsevier)

In addition, the pulsed laser deposition technique has been widely used for the deposition of multilayers with different compounds. Pulse laser deposition is a versatile technique for the growth of thin films that allows the deposition of various materials such as ferroelectrics, polymers, semiconductors, or metal oxide materials. Pulse laser deposition is more versatile than other film growth techniques because the choice of laser wavelength and other growth parameters allows one to control the size, shape, and crystalline phase of films and clusters. For each ferrite composition, the best conditions for the growth of well-crystallized films may be determined. Nanometric or micrometric spinel ferrite films can be obtained by modifying the deposition parameters. Furthermore, these processing modifications can tune the magnetic properties since differences in the distribution of cations on the surface and in the mass can be obtained [63, 64].

6.5 Sintering

The production of bulk ceramic materials (including ferrites) goes through a processing step that aims to achieve microstructures with high density, that is, that have a minimum or practically non-existent porosity, such that their intrinsic properties prevail in their entirety. This processing step is known as sintering. It can be described as a process in which a compacted powder of a crystalline or non-crystalline solid undergoes a firing process, during which the pores between the initial particles are eliminated, resulting in a retraction of the body, combined with a growth of its grains (coalescence) and an intense union between adjacent particles [65].

The driving force for sintering is the reduction of the free energy of the system, achieved by decreasing the specific surface due to the growth of the contact areas of the particle, the decline of the pore volume and/or spheroidization of these, the elimination of the concentrations of non-equilibrium network defects (point defects, discrepancies) in the powder mass, and the removal of non-equilibrium states due to mutual solid solubility or chemical reactivity, in the case of multicomponent systems [1, 3, 66, 67].

For sintering to occur, a mechanism for transporting material is required, in addition to a source of energy (heat) to activate the process and sustain this transport. Possible pathways for matter transfer are evaporation and condensation, surface diffusion, volumetric diffusion, grain contour diffusion, or viscous flow. The latter is predominant in vitreous systems [68].

Sintering can occur according to one of the two main mechanisms: sintering by solid phase and sintering by the liquid phase. In sintering in the solid-state, there is no fusion or liquid phase formation, and the transport of matter occurs by diffusion. In this case, sintering can occur with or without solid solubility of the components. On the other hand, in sintering with the liquid phase, one of the present phases melts and the transport of matter occurs by capillary movement of the formed liquid and diffusion through this medium. The amount of liquid phase can reach 30 vol.% at the sintering temperature, provided that the liquid has a good wettability concerning the solid and that there is a good solubility of the solids in the liquid.

All the consolidation methods generally used in micrometric ferrite processing can also be used for nanocrystalline materials. However, because of the smaller size, some special precautions need to be taken. Successful consolidation of nanocrystalline powders is a non-trivial problem since fully dense materials should be produced while simultaneously retaining the nanosized grains without coarsening. Conventional consolidation of powders to full density through processes such as uniaxial pressing and sintering may not be as efficient in these cases since the energy supplied is not sufficient to reduce the porosity of the materials. On the other hand, hot pressing or hot isostatic pressing requires the use of high pressures and elevated temperatures for extended periods to achieve full densification [69].

Unfortunately, this results in significant coarsening of the nanosized grains, and consequently, the benefits of nanostructure processing are lost. On the other hand, retention of nanostructures requires the use of low consolidation temperatures, and it is difficult to achieve full interparticle bonding at these low temperatures. Therefore, to sinter nanosized ferrites, novel and innovative methods of consolidating nanocrystalline powders are required. This method includes spark plasma sintering (SPS), microwave sintering, cold sintering, fast firing, and hot isostatic pressing (HIP).

By utilizing the combination of high temperature, pressure, and/or electric discharges, one can achieve a particular density at a lower pressure when compared to cold isostatic pressing or at a lower temperature when compared to conventional sintering. The results of some of these investigations [35, 70–73] show that because of the increased diffusivity in nanocrystalline materials, sintering (densification) takes place at temperatures much lower than in coarse-grained materials. This is likely to reduce grain growth. Moreover, nanocrystalline materials have been shown to exhibit increased strength/hardness, enhanced diffusivity, reduced density, higher electrical resistivity, increased specific heat, a higher coefficient of thermal expansion, lower thermal conductivity, and superior soft magnetic properties in comparison with their coarse-grained counterparts. Thus, it becomes crucial to obtain fully dense materials while simultaneously retaining nanometer-sized grains to demonstrate the improvement in properties due to nanostructure processing unambiguously.

6.5.1 Spark Plasma Sintering (SPS)

The spark plasma sintering technique has stood out in recent years [74] compared to other methods due to its advantages, that include: lower sintering temperatures, short landing time, and the ability to densify sturdy materials. This technique allows using temperatures and sintering times lower than those used in the sintering using conventional methods, which renders it as an effective technique for obtaining dense nanocrystalline materials with minimal grain growth.

In summary, the spark plasma sintering technique differs from other sintering techniques by how the samples are heated. In this method, the powder is pressed uniaxially in a graphite mold, and high electric currents are applied directly to the mold promoting high heating rates in addition to the simultaneous application of pressure. The powder is heated by the discharge between the particles and by the electric current through the graphite mold. Due to the discharges, the surfaces of the particles are activated, and a self-heating phenomenon is generated between them and thus, the transfer of heat and mass occurs quickly. The process of spark plasma sintering technique is characterized by high thermal efficiency due to the direct heating of the graphite mold and the powder to be sintered by the passage of high alternating currents [75]. Figure 6.23 shows the typical configuration of the SPS

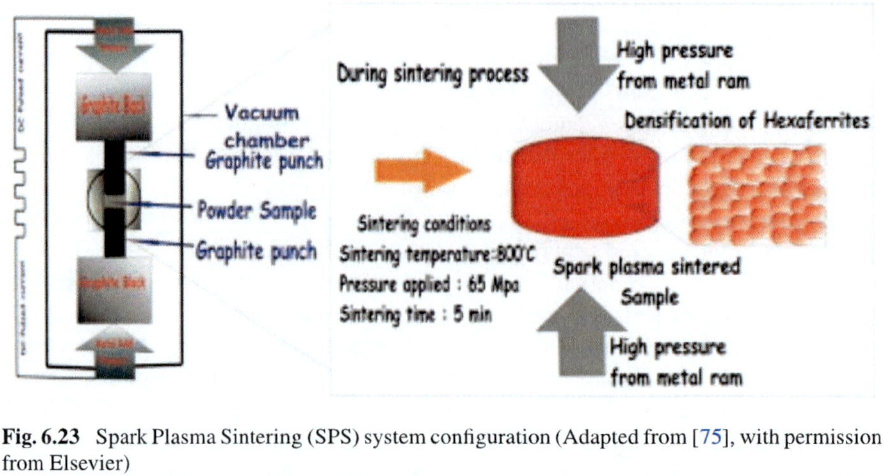

Fig. 6.23 Spark Plasma Sintering (SPS) system configuration (Adapted from [75], with permission from Elsevier)

system.

In the conventional hot-pressing process, for example, the main factor that promotes densification is the heating generated by electrical resistance and the plastic flow of materials due to the application of pressure. In the case of the plasma sintering process, in addition to these factors, there is also the direct application of electrical current. The alternating current pulse applied directly to the graphite mold and the sample it can generate: (a) spark plasma, (b) impact pressure from plasma generation, (c) Joule heating, and (d) electric field diffusion [76]. Typical spark plasma sintering processing parameters include:

a. pressures applied between 50 and 100 MPa (when the mold is graphite, as the pressure may depend on the material of the mold);
b. heating rates up to 600 ºC/min;
c. pulse duration of ~ 10 ms with cycles of 2–5 ms; and
d. maximum pulse parameters of 10,000 A and 10 V. The start of sintering depends on the initial particle size, which limits the beginning of the contraction. It is usually difficult to determine when the sintering process has been completed.

However, the displacement of the piston (applying pressure) can help to determine the end of SPS sintering [77]. The spark plasma sintering process offers several advantages over conventional means, such as hot pressing and hot isostatic pressing, including ease of operation, high sintering speed, high reproducibility, safety, and reliability, in addition to lower energy consumption, which means SPS is usually one-third to one-fifth more economical than the processes presented above [78]. The spark plasma sintering technique is also widely used to manufacture spinel ferrites with minimal grain growth and high density. This process offers a resistive heating mechanism due to the high amplitude of the pulsed direct current and, therefore, causes a significant increase in temperature within the powder particles in short time intervals. This effect significantly reduces grain growth. When comparing SPS with conventional sintering, it is realized that conventional sintering is a long-lasting process, often resulting in excessive grain growth during sintering. Millot and collaborators [74] produced nanocrystalline $CoFe_2O_4$. A pure spinel phase has been obtained, with no grain growth observed, and with significant densification of 91% when spark plasma sintering was used. The SPS method seems to be a promising process to produce dense nanostructured ferrites. Figure 6.24 shows the comparison SPS (900, with a heating rate of 100 ºC/min in 14 min) and conventional sintering (1350 °C, with a heating rate of 5 °C/min in 4 h). It is possible to observe that the same density has been achieved (about 91%) by both sintering methods. However, because of the shorter process time (14 min instead of more than 4 h) and limited grain growth, SPS is more interesting from an economic point of view.

In addition to the sintering of nanometric materials, the ability of spark plasma sintering to reduce the sintering temperature is fascinating, especially in some ferrites that undergo phase decomposition with increasing temperature. Magnetite can be sintered by spark plasma sintering in a dense, high-crystalline bulk at 750 °C for 15 min, with a grain size of 150 nm [79] without oxidation to maghemite or hematite. The reductive atmosphere of the sintering spark plasma prevents oxidation. Another

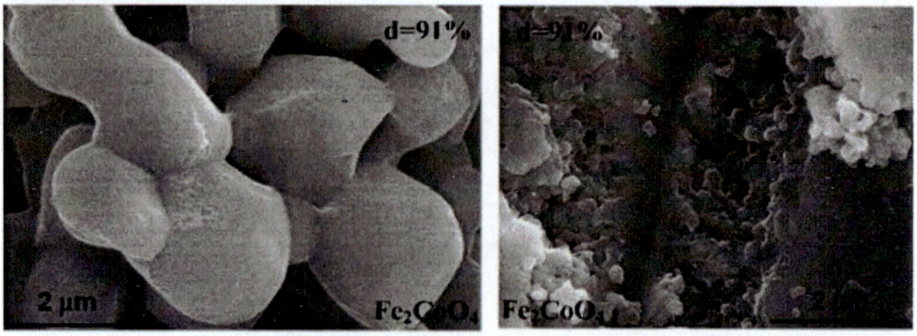

(a) (b)

Fig. 6.24 Comparison of a conventional sintering (**a**) and a SPS of the same powder of $CoFe_2O_4$ (**b**) (Adapted from [74], with permission from Elsevier)

example is BFO, because it slowly decomposes into $Bi_2Fe_4O_9$ above 675 °C, and then quickly transforms to $Bi_2Fe_4O_9$ above 830 °C. When BFO ceramics are fabricated by SPS, they exhibit a high bulk density of up to 96% theoretical density and do not decompose into another phase. Therefore, their ferroelectric, piezoelectric, and magnetic properties are improved concerning those of the conventionally sintered ones. It is important to emphasize that extra care must be taken during sintering by spark plasma sintering. This process may not allow getting the desired spinel ferrite systematically from whatever starting powders compared to conventional sintering. Sometimes the sintering process is so fast that there is no time for equilibrium to be reached and phase formation occurs. On the other hand, the metallic ions may not have completely dissolved in the spinel structure or may have segregated as another oxide [74]. Despite the excellent results obtained by this technique, there are some limitations, such as the need for a graphite mold, limited shapes, very high cost, and less sintering efficiency for non-metallic materials.

6.5.2 *Microwave Sintering*

In conventional thermal processing, the energy is transferred to the material through processes of conduction, convection, and radiation to the surface of the material. Then, through driving processes and convection the heat is conducted into the interior of the material. In microwave processing, the energy is deposited directly inside the material through molecular and atomic interaction processes with the electromagnetic field, in the same way as explained in Chapter 5. This difference in the form of heating provides several advantages to the processing of materials using microwaves with respect to heating techniques conventional. As heating takes place through interaction with the electromagnetic field, it occurs throughout the entire volume of the material, which makes it possible to work with high heating rates and

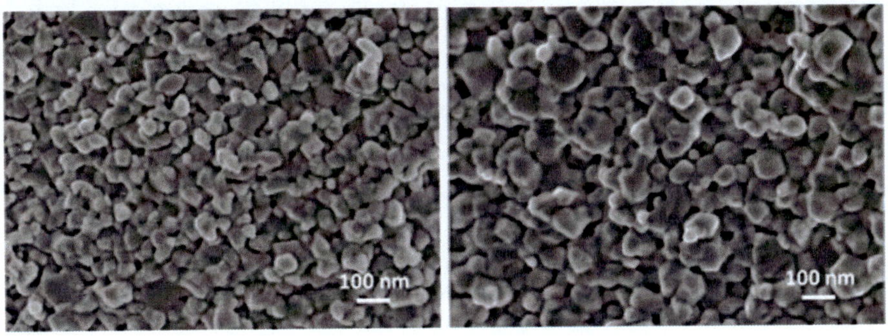

Fig. 6.25 Microstructure of a nanometric $CoFe_2O_4$ and sintered by microwaves at 850 °C (left) and 900 °C (right) (Reprinted from [80], with permission from Elsevier)

reduced processing cycles. Selective heating of the material to be processed also allows for a reduction in energy costs.

The materials obtained after sintering have a significant improvement in the microstructural uniformity of the ceramic bodies, better product yield, energy savings, and reduced manufacturing cost [35]. In ferrites, as in all ceramic materials, the high temperatures required to fully densify ceramic powders result in large grain sizes. This phenomenon occurs due to Ostwald's ripening when traditional sintering techniques are used. That is why it is so difficult to obtain dense materials with nanometric and submicrometric grain sizes using conventional techniques. Microwave sintering has recently been used as an efficient technique to hinder grain growth and also to produce a homogeneous microstructure. Other benefits of microwave sintering over conventional sintering techniques include the reduction of cracks and thermal stress, increased strength, and reduced contamination.

The microstructure of a nanometric $CoFe_2O_4$ and sintered by microwaves are shown in Fig. 6.25. The authors found a fine microstructure with the grain size well below 100 nm and high homogeneity at both microwave sintering temperatures [80].

References

1. German RM (1984) Powder metallurgy science. Metal Powder Industries Fed, Princeton
2. Moreno R (2019) Better ceramics through colloid chemistry. J Eur Ceram Soc. https://doi.org/10.1016/j.jeurceramsoc.2019.10.014
3. Setz LFG, da Silva AC (2019) O processamento cerâmico sem mistério. Blucher, São Paulo
4. Moreno R (2012) Colloidal processing of ceramics and composites. Adv Appl Ceram. https://doi.org/10.1179/1743676111Y.0000000075
5. Moreno Botella R (2005) Reología de suspensiones cerámicas. Consejo Superior de Investigaciones Científicas
6. Yu F, Chen Y, Liang X et al (2017) Dispersion stability of thermal nanofluids. Prog Nat Sci Mater Int 27:531–542

7. Ortega FS, Pandolfelli VC, Rodrigues JA, de Souza DPF (1997) Aspectos da Reologia e da Estabilidade de Suspensões Cerâmicas. Parte III: Mecanismo de Estabilização Eletroestérica de Suspensões com Alumina. Cerâmica. https://doi.org/10.1590/s0366-69131997000300003
8. Parks GA, De Bruyn PL (1962) The zero point of charge of oxides. J Phys Chem. https://doi.org/10.1021/j100812a002
9. Martin Cabañas B, Leclercq S, Barboux P et al (2011) Sorption of nickel and cobalt ions onto cobalt and nickel ferrites. J Colloid Interface Sci. https://doi.org/10.1016/j.jcis.2011.04.082
10. Kosmulski M (2009) Surface charging and points of zero charge. CRC Press, Boca Raton
11. Vidojkovic SM, Rakin MP (2017) Surface properties of magnetite in high temperature aqueous electrolyte solutions: a review. Adv. Colloid Interface Sci 245:108–129
12. Milanovic M, Stijepovic I, Pavlovic V, Srdic VV (2016) Functionalization of zinc ferrite nanoparticles: influence of modification procedure on colloidal stability. Process Appl Ceram. https://doi.org/10.2298/PAC1604287M
13. Tang W, Su Y, Li Q et al (2013) Superparamagnetic magnesium ferrite nanoadsorbent for effective arsenic (III, V) removal and easy magnetic separation. Water Res. https://doi.org/10.1016/j.watres.2013.04.023
14. Campos AFC, Aquino R, Tourinho FA et al (2013) Influence of the spatial confinement at nanoscale on the structural surface charging in magnetic nanocolloids. Eur Phys J E. https://doi.org/10.1140/epje/i2013-13042-y
15. Neimark AV (2004) Colloids and surfaces A: physicochemical and engineering aspects—foreword. Colloids Surf A: Physicochem Eng Asp. https://doi.org/10.1016/S0927-7757(04)00192-X.
16. Cesconeto FR, Rodrigues Neto JB, Arcaro S et al (2014) Reologia de suspensões de precursor vitrocerâmico do sistema LiO_2-ZrO_2-SiO_2-Al_2O_3. Cerâmica 60:149–153. https://doi.org/10.1590/S0366-69132014000100020
17. Arcaro S, Oliveira APN, Gutiérrez-Gonzalez CF et al (2017) LZS/Al_2O_3 nanostructured composites obtained by colloidal processing and spark plasma sintering. J Eur Ceram Soc 37. https://doi.org/10.1016/j.jeurceramsoc.2017.03.023
18. Arcaro S, Wermuth TB, Zampiva RYS et al (2019) Li_2O-ZrO_2-SiO_2/Al_2O_3 nanostructured composites for microelectronics applications. J Eur Ceram Soc 39:491–498. https://doi.org/10.1016/j.jeurceramsoc.2018.09.033
19. Arcaro S, Oliveira APN, Gutiérrez-Gonzalez CF et al (2017) LZS/Al2O3 nanostructured composites obtained by colloidal processing and spark plasma sintering. J Eur Ceram Soc. https://doi.org/10.1016/j.jeurceramsoc.2017.03.023
20. Arcaro S, Isabel Nieto M, Rodrigues Neto JB et al (2016) Al_2O_3 nanoparticulate LZS glass-ceramic matrix composites for production of multilayered materials. J Am Ceram Soc. https://doi.org/10.1111/jace.14418
21. Arcaro S, Goulart De Oliveira Maia B, Souza MTMT et al (2016) Thermal insulating foams produced from glass waste and banana leaves. Mater Res 19:1064–1069
22. Mori T, Imazeki H, Yamamoto K, Tsuchiya K (2017) Effects of a slurry's particle dispersion and flocculation state on spray-dried granule morphology. Ceram Int. https://doi.org/10.1016/j.ceramint.2017.05.165
23. Yan H, Zhang Y, Wang Y et al (2019) Morphology-selected synthesis of copper ferrite via spray drying with excellent sodium storage properties. Ceram Int. https://doi.org/10.1016/j.ceramint.2019.07.066
24. Nireesha G, Divya L, Sowmya C et al (2013) Lyophilization/freeze drying—an review. Int J Nov Trends Pharm Sci
25. Kogias G, Tsakaloudi V, Van Der Valk P, Zaspalis V (2012) Improvement of the properties of MnZn ferrite power cores through improvements on the microstructure of the compacts. J Magn Magn Mater. https://doi.org/10.1016/j.jmmm.2011.07.055
26. de Oliveira APN, Hotza D (2015) Tecnologia de fabricação de revestimentos cerâmicos. Editora da UFSC
27. Arcaro S, Neto JBR, Cesconeto FR et al (2014) Processing of silicas formed by slip casting. Mater Sci Forum 775–776: 525–528

28. Le Ferrand H (2021) Magnetic slip casting for dense and textured ceramics: a review of current achievements and issues. J Eur Ceram Soc 41:24–37
29. Deville S (2008) Freeze-casting of porous ceramics: a review of current achievements and issues. Adv Eng Mater 10:155–169
30. Dos Santos AL, Marçal RLSB, De Campos JB et al (2018) Magnetic filter produced by $ZnFe_2O_4$ nanoparticles using freeze casting. J Mater Res Technol 7:350–355. https://doi.org/10.1016/j.jmrt.2018.04.012
31. Sepulveda P (1997) Gelcasting foams for porous ceramics. Am Ceram Soc Bull 76: 61–65
32. Stochero NP, de Moraes EG, Moreira AC et al (2020) Ceramic shell foams produced by direct foaming and gelcasting of proteins: permeability and microstructural characterization by X-ray microtomography. J Eur Ceram Soc 40:4224–4231. https://doi.org/10.1016/j.jeurceramsoc.2020.05.036
33. Pozzobom IEF, de Souza M, Neto JBR et al (2014) Production of Li_2O-ZrO_2-SiO_2-Al_2O_3 (LZSA) glass-ceramic foams by aeration and polymerization of suspension. Mater Sci Forum 775–776:529–533. https://doi.org/10.4028/www.scientific.net/MSF.775-776.529
34. Eunice I, Pozzobom F, De Moraes GG et al (2015) Glass-ceramics foam for hydrogen production. Chem Eng Trans 43:1789–1794. https://doi.org/10.3303/CET1543299
35. Liu L, Yan Y, Ngo KDT, Lu GQ (2017) NiCuZn ferrite cores by gelcasting: processing and properties. IEEE Trans Ind Appl 53:5728–5733. https://doi.org/10.1109/TIA.2017.2746038
36. Liu L, Ngo KDT, Lu GQ (2021) Effects of SiO_2 inclusions on sintering and permeability of NiCuZn ferrite for additive manufacturing of power magnets. J Eur Ceram Soc 41:466–471. https://doi.org/10.1016/j.jeurceramsoc.2020.08.004
37. Smith DL, Hoffman DW (1996) Thin-film deposition: principles and practice. Phys Today 49:60–62. https://doi.org/10.1063/1.2807590
38. IUPAC (2009) IUPAC compendium of chemical terminology. IUPAC
39. Ter M-S (1994) Thin films including layers: terminology in relation to their preparation and characterization (IUPAC Recommendations 1994). Pure Appl Chem 66:1667–1738. https://doi.org/10.1351/pac199466081667
40. Hotza D, Greil P (1995) Aqueous tape casting of ceramic powders. Mater Sci Eng A-Structural Mater Prop Microstruct Process 202:206–217. https://doi.org/10.1016/0921-5093(95)09785-6
41. Moreno R (1992) The role of slip additives in tape casting technology. II: binders and plasticizers. Am Ceram Soc Bull 71:1647–1657
42. Jabbari M, Bulatova R, Tok AIY et al (2016) Ceramic tape casting: a review of current methods and trends with emphasis on rheological behaviour and flow analysis. Mater Sci. Eng. B Solid-State Mater. Adv. Technol. 212:39–61
43. Hamaker HC (1940) Formation of a deposit by electrophoresis. Trans Faraday Soc 35:279. https://doi.org/10.1039/tf9403500279
44. Ferrari B, Moreno R (2010) EPD kinetics: a review. J Eur Ceram Soc 30:1069–1078
45. Hashi S, Takada N, Nishimura K et al (2005) Fabrication technique for over 10-μm-thick ferrite particulate film at room temperature. IEEE Trans Magn 41:3487–3489. https://doi.org/10.1109/TMAG.2005.854912
46. Hashi S, Yabukami S, Maeda A et al (2007) High-frequency magnetic properties of ferrite particulate films. J Magn Magn Mater 316:465–467. https://doi.org/10.1016/J.JMMM.2007.03.172
47. Bornside DE, Macosko CW, Scriven LE (1987) Modeling of spin coating. J Imaging Technol 13:122–130
48. Tudorache F, Popa PD, Dobromir M, Iacomi F (2013) Studies on the structure and gas sensing properties of nickel-cobalt ferrite thin films prepared by spin coating. Mater Sci Eng B Solid-State Mater Adv Technol. https://doi.org/10.1016/j.mseb.2013.03.019
49. Ben KS, Gassoumi M, Ben DA et al (2019) The effect of the cobalt ferrites nanoparticles ($CoFe_2O_4$) on the porous silicon deposited by spin coating. Alexandria Eng J. https://doi.org/10.1016/j.aej.2019.12.031
50. Quandt N, Syrowatka F, Roth R et al (2019) Spin coating of epitaxial $BaTiO_3/CoFe_2O_4$ thin films on single crystalline (001)-$SrTiO_3$. Mater Chem Phys 229:453–459. https://doi.org/10.1016/j.matchemphys.2019.03.012

51. Borlaf M, Colomer MT, Moreno R, Ortiz AL (2014) Rare earth-doped TiO_2 nanocrystalline thin films: preparation and thermal stability. J Eur Ceram Soc. https://doi.org/10.1016/j.jeurceramsoc.2014.07.008
52. Illa R,Ješko R, Silber R et al (2019) Structural, magnetic, optical, and magneto-optical properties of $CoFe_2O_4$ thin films fabricated by a chemical approach. Mater Res Bull 117:96–102. https://doi.org/10.1016/j.materresbull.2019.05.002
53. Kotsikau D, Ivanovskaya M, Pankov V, Fedotova Y (2015) Structure and magnetic properties of manganese-zinc-ferrites prepared by spray pyrolysis method. Solid State Sci 39:69–73. https://doi.org/10.1016/j.solidstatesciences.2014.11.013
54. Sarıtaş S, Şakar BC, Turgut E et al (2021) Cobalt metal doped magnesium ferrite and zinc ferrite thin films grown by Spray Pyrolysis. Mater Today Proc. https://doi.org/10.1016/j.matpr.2021.03.284
55. Hwang TY, Lee J, Lim HR et al (2017) Synthesis and magnetic properties of La^{3+}-Co^{2+} substituted strontium ferrite particles using modified spray pyrolysis–calcination method. Ceram Int 43:3879–3884. https://doi.org/10.1016/j.ceramint.2016.12.051
56. Charishma VD, Shastrimath V, Bairy R et al (2019) Structural, morphological and optical properties of barium doped bismuth ferrite thin films deposited by spray pyrolysis. Mater Today Proc 35:440–444. https://doi.org/10.1016/j.matpr.2020.02.950
57. Kshamitha D, Manjunath V, Dastagiri S et al (2020) Influence of annealing on structural and morphological of spray pyrolysis deposited $CoFe_2O_4$ thin films. South African J Chem Eng 33:51–57. https://doi.org/10.1016/j.sajce.2020.04.004
58. Ran M, Yu Z, Sun K et al (2021) Effects of aqueous ethanol solutions on the structural and magnetic properties of NiZn ferrite thin films prepared by spin-spray deposition. Ceram Int. https://doi.org/10.1016/j.ceramint.2021.02.119
59. Baraliya JD, Joshi HH (2014) Spectroscopy investigation of nanometric cobalt ferrite synthesized by different techniques. Vib Spectrosc 74:75–80. https://doi.org/10.1016/J.VIBSPEC.2014.07.013
60. Madake SB, Hattali MR, Thorat JB et al (2021) Chemiresistive gas sensing properties of copper substituted zinc ferrite thin films deposited by spray pyrolysis. J Electron Mater 50:2460–2465. https://doi.org/10.1007/s11664-021-08783-1
61. Shen L, Althammer M, Pachauri N et al (2014) Epitaxial growth of spinel cobalt ferrite films on $MgAl_2O_4$ substrates by direct liquid injection chemical vapor deposition. J Cryst Growth 390:61–66. https://doi.org/10.1016/j.jcrysgro.2013.12.012
62. Fietzke F, Zywitzki O (2017) Structure and properties of magnetron-sputtered manganese ferrite films. Thin Solid Films 644:138–145. https://doi.org/10.1016/j.tsf.2017.10.034
63. Sánchez-Arenillas M, Oujja M, Moutinho F et al (2019) Bulk and surface characterisation of micrometer-thick cobalt ferrite films grown by IR PLD. Appl Surf Sci 470:917–922. https://doi.org/10.1016/J.APSUSC.2018.11.152
64. Oujja M, Martín-García L, Rebollar E et al (2018) Effect of wavelength, deposition temperature and substrate type on cobalt ferrite thin films grown by pulsed laser deposition. Appl Surf Sci 452:19–31. https://doi.org/10.1016/j.apsusc.2018.05.012
65. Kingery WD, Berg M (1955) Study of the initial stages of sintering solids by viscous flow, evaporation-condensation, and self-diffusion. J Appl Phys 26:1205–1212. https://doi.org/10.1063/1.1721874
66. Reed JS (1995) Principles of ceramics processing. Wiley, New York
67. German RM (1983) Fundamentals of sintering—ceramics and glasses handbook, vol 4. pp 260–269
68. Thümmler F, Oberacker R (1993) An introduction to powder metallurgy. Institute of Materials, London
69. Suryanarayana C, Koch CC (2000) Nanocrystalline materials-Current research and future directions. Springer
70. Venturini J, Wermuth TB, Machado MC et al (2019) The influence of solvent composition in the sol-gel synthesis of cobalt ferrite ($CoFe_2O$): a route to tuning its magnetic and mechanical properties. J Eur Ceram Soc 39. https://doi.org/10.1016/j.jeurceramsoc.2019.01.030

71. Pribošič I, Makovec D, Drofenik M (2005) Chemical synthesis of $KNbO_3$ and $KNbO_3$-$BaTiO_3$ ceramics. J Eur Ceram Soc. https://doi.org/10.1016/j.jeurceramsoc.2005.03.128
72. Yuan Y, Wang B, Wang C et al (2017) Effects of $CoFe_2O_4$ electrode microstructure on the sensing properties for mixed potential NH_3 sensor. Sensors Actuators, B Chem. https://doi.org/10.1016/j.snb.2016.07.171
73. Caldeira LE, Guaglianoni WC, Venturini J et al (2020) Sintering-dependent mechanical and magnetic properties of spinel cobalt ferrite ($CoFe_2O_4$) ceramics prepared via sol-gel synthesis. Ceram Int 46:2465–2472. https://doi.org/10.1016/j.ceramint.2019.09.240
74. Millot N, Le Gallet S, Aymes D et al (2007) Spark plasma sintering of cobalt ferrite nanopowders prepared by coprecipitation and hydrothermal synthesis. J Eur Ceram Soc 27:921–926. https://doi.org/10.1016/j.jeurceramsoc.2006.04.141
75. Hungría T, Galy J, Castro A (2009) Spark plasma sintering as a useful technique to the nanostructuration of piezo-ferroelectric materials. Adv Eng Mater 11:615–631. https://doi.org/10.1002/adem.200900052
76. Breitwieser R, Acevedo U, Ammar S, Valenzuela R (2017) Ferrite nanostructures consolidated by spark plasma sintering (SPS). In: Nanostructured materials—fabrication to applications. InTech, Rijeka
77. Ishiyama, M (1993) Plasma Activated Sintering (PAS) System. Proc 1993 Powder Met World Congr 931–934
78. Omori M (2000) Sintering, consolidation, reaction and crystal growth by the spark plasma system (SPS). Mater Sci Eng A 287:183–188. https://doi.org/10.1016/s0921-5093(00)00773-5
79. Gaudisson T, Vázquez-Victorio G, Bañobre-López M et al (2014) The Verwey transition in nanostructured magnetite produced by a combination of chimie douce and spark plasma sintering. J Appl Phys 115:17E117
80. Fariñas JC, Moreno R, Pérez A et al (2018) Microwave-assisted solution synthesis, microwave sintering and magnetic properties of cobalt ferrite. J Eur Ceram Soc 38:2360–2368. https://doi.org/10.1016/J.JEURCERAMSOC.2017.12.052

Chapter 7
Current and Incipient Applications

Ferrites have attracted significant scientific and technological interest due to the possibility of specific applications through their chemical manipulation concerning size, morphology, magnetic behavior, and chemical composition.

In this chapter, we describe the main applications of these materials. These applications permeate the biomedical, electrical devices, and water treatment areas. In this sense, the main advances of spinel ferrites in applications in magnetic hyperthermia, magnetic separation, drug delivery, magnetic resonance image, EMI filters, sensors, and adsorption are shown.

7.1 Biomedical Applications

In the biomedical field, the great interest in ferrites is focused on contrast agents in nuclear magnetic resonance imaging, vectorization, drug delivery, destruction of tumor cells by magnetic hyperthermia, magnetic separation of nucleic acids, and the reduction of viral loads [1–3]. In the vast majority of cases, these particles are found in nanometric sizes. These magnetic nanoparticles are generally formed by magnetite (Fe_3O_4), cobalt ferrite ($CoFe_2O_4$), and nickel ferrite ($NiFe_2O_4$), among other oxides with spinel structure. The main advantage of using ferrite nanoparticles in biomedical applications involves their ability to interact with the biological entities of interest due to the relative sizes of the nanoparticles, smaller than cells (10–100 μm) and comparable to viruses (20–450 nm), proteins (5–50 nm) or genes (2 nm wide and 10–100 nm long). Furthermore, nanoferrites respond to an external magnetic field gradient, and they have a large surface area/volume ratio which allows their surface to be modified for specific biomedical applications. Moreover, the surface can be modified to improve stability, avoid aggregation of particles, and improve biocompatibility and dispersion, thus increasing circulation time, while simultaneously mitigating its cytotoxicity [4, 5].

S. Arcaro and J. Venturini, *Modern Ferrites in Engineering*, Topics in Mining, Metallurgy and Materials Engineering, https://doi.org/10.1007/978-3-030-78988-6_7

Although they are very useful and with great application potential, nanoferrites also present obstacles. The size of these particles can cause some stability problems: overtime, they tend to agglomerate, and if not covered with a protective layer, then they may oxidize.

Biomedical applications of magnetic nanoparticles (MNPs) involve both administration within the organism (in vivo) and external to the organism (in vitro). It is very important to know whether the administration of these nanomaterials is safe or whether they are toxic to healthy cells, producing long-term health problems. These concerns have gained a lot of attention from researchers around the world, who are looking to develop versatile nanomaterials with a balance between application and toxicity.

7.1.1 Magnetic Hyperthermia

Cancer is the main public health problem globally, and it is among the top four causes of premature death in most countries. Cancer incidence and mortality have been increasing worldwide, partly due to aging, population growth, and the change in distribution and prevalence of risk factors, especially those associated with socioeconomic development. Currently, the conventional modalities used in the treatment of cancer are surgery, chemotherapy, and radiotherapy. However, none of these alone can completely eradicate malignant tumor tissue. Surgical treatment offers the best chance of cure. Both chemotherapy and radiotherapy are aggressive treatments and cause several side effects. For this reason, several efforts are directed toward developing more effective cancer treatment and drugs with fewer side effects and lower costs, and, at the same time, improving in the patient's quality of life. However, few new features have been added to conventional therapy. Hyperthermia is a proposal for cancer treatment in which the tumor cells are affected by the rise in local temperature to temperatures in the range of 43–45 °C. Under these conditions, cancer cells are more sensitive to the cytotoxic effects of heat, and, therefore, hyperthermia leads to cell death, preferably with high selectivity of tumor cells over healthy cells. Different approaches, such as laser, microwave, radiofrequency, and ultrasound, have been used to apply hyperthermia to cancerous regions. However, as the energy source in these techniques is external, most of the energy is dissipated in healthy cells along the radiation path. This lack of selectivity causes numerous side effects [8]. In this context, magnetic nanoparticles—more precisely ferrites—have been researched as hyperthermia agents to offer a localized treatment. The increase in temperature required by hyperthermia can be achieved, among other methods, by the use of magnetic nanoparticles [6]. Magnetic nanoparticles under a magnetic field randomly rotate their direction of magnetization between parallel and antiparallel orientations. Such behavior allows the transfer of magnetic energy to the particles in the form of heat, a property that can be used in vivo to increase the temperature of the tumor in the tissues and thermally destroy the pathological cells.

In magnetic hyperthermia, two mechanisms are responsible for heat generation from magnetic nanoparticles: magnetic relaxation, which can be Brown or Néel, and loss due to hysteresis. The effect of heating nanoparticles with magnetic multidomain occurs due to hysteresis losses. Magnetic relaxation occurs mainly in single-domain nanoparticles, such as superparamagnetic nanoparticles [7]. The release of heat by Brown's relaxation is related to the rotation of the particles and shocks between them, while Néel's relaxation is related to the rapid change of direction of the magnetic moments with regard to the crystalline structure of the material. Small magnetic nanoparticles are more appropriate as they require less energy to rotate their magnetic moments. Thus, the ideal size of nanoparticles should be around 15 nm to optimize the balance between the Néel and Brown contributions [8]. Heat is dissipated in the magnetic nanoparticles due to the delay in relaxing the nanoparticles. Therefore, the value of heat dissipation depends on the frequency and amplitude of the applied magnetic field, in addition to the magnetic properties of the nanoparticles. The transformation of magnetic energy into thermal energy is quantified through the specific absorption rate (SAR). This value depends on the size, shape, and magnetic properties of the nanoparticles, the frequency and amplitude of the applied magnetic field, the characteristics of the coating material, and the functionalization of the surface. Nanoparticles with a high SAR value are desirable for clinical applications, as they allow the use of a smaller dosage of nanoparticle solutions, thus decreasing the likelihood of deleterious side effects [4].

During overheating in the region of cancer cells, there is a change in the synthesis of DNA and RNA, causing denaturation of enzymes, impairment of the functions of mitochondria, and destabilization of cell membranes that lack oxygen. Under these conditions, cancer cells are more sensitive to the cytotoxic effect of heat, and, therefore, hyperthermia causes the selective death of tumor cells over healthy cells [7]. Subsequently, the body itself processes the dead cells and absorbs their structures. The temperature increase required by hyperthermia can be achieved, among other methods, by the use of magnetic nanoparticles, a process known as magnetohyperthermia [3, 6, 9–11].

Hyperthermia can be classified according to the increase in temperature during treatment:

i. Thermal ablation (>46 °C);
ii. Moderate hyperthermia (41–46 °C), mainly used to treat cancer;
iii. Diathermy (<41 °C), which may be used to treat rheumatic diseases.

Above 46 °C, necrosis of healthy cells occurs. Moderate hyperthermia.[7].

The three types of treatment may be performed depending on the location of the disease in which the effect is applied. Local hyperthermia consists of heating a specific area of tissue, such as a tumor. In regional hyperthermia, a large area of tissue, such as an entire organ, is heated. On the other hand, whole-body hyperthermia involves heating the entire body for the treatment of metastatic cancer.

Several studies show the success of the application of nanoparticles in the treatment of experimental tumors [3, 12–16]. Most of the studies show a significant delay in the development and growth of cancer, and some showed complete remission of the tumor mass. On the other hand, the combination of magnetohyperthermia and chemotherapy can also improve the effectiveness of antineoplastic drugs without increasing their cytotoxic potential. The thermal action increases the permeability of the cell membrane, leaving the cell more vulnerable to attack by chemical substances. Minimizing damage to surrounding healthy tissues makes magnetohyperthermia a promising technique for the treatment of various cancers. Phase III clinical experiments prove that when hyperthermia was associated with radiotherapy, it improved the response rate from 41 to 59% in the case of breast cancer; it caused an increase in local control and survival from 37 to 56% and 20% to 37%, respectively, for cervical cancer. For lung cancer, the application of hyperthermia in radiotherapy increased the average survival from 7–8 months to 18 months [17]. Although many superparamagnetic nanoparticle compositions of spinel ferrites (MFe_2O_3, where M = Ni, or Co, or Fe, and others) have been studied, Fe_3O_4 is currently approved for in vivo applications by the FDA (Food and Drug Administration). Also, magnetite particles present the most significant advances in studies for applications in medicine.

The interaction between the particles does not cause agglomeration (thus avoiding thrombosis) and has shown effectiveness for in vivo application. The main obstacle toward the utilization of MNPs is making the nanoparticles reach a particular location in the body, being stable in body fluids, and having reduced toxicity [18]. Therefore, the surface of these particles may be modified by coating with organic polymers (dextran, PEG, PVP, PVA, and dendrimers), metals (Au, Gd, and Pt), oxides (SiO_2, Al_2O_3, TiO_2), fatty acids (such as oleic acid), or amino acids. The latter are suitable for further functionalization by binding various bioactive molecules, such as biotin, avidin, carboxyl groups, carbodiimide, antibodies, hormones, folic acid, albumin proteins, among others. In addition to the superficial modification promoting the more specific interaction of the particle with the tissue, cell, or molecule, the surface covering also contributes to the increase in biocompatibility [19–21].

The heating efficiency of nanostructured spinel ferrites is dependent on the number of nanoparticles in suspension, as well as their means of stabilization. In addition, it is possible to achieve improved heating efficiency from the control of the stability. The surface passivation offers colloidal stability, as well as prevents their aggregation under the AC magnetic field.

The heating efficiency of colloidally stable ferrites was investigated under the application of an AC magnetic field using different media (water, phosphate-buffered saline, cell culture medium) to guarantee their properties in in vivo conditions. Figure 7.1 shows the graphs of temperature vs. time of Fe_3O_4 nanoparticles coated with proteins at a concentration of 1 mg/ml of Fe in different media under an AC magnetic field of 423 Oe and frequency of 300 kHz. The time required to reach 42–43 °C was around 20 min. SAR values were 69 W/g to 94.5 W/g [22].

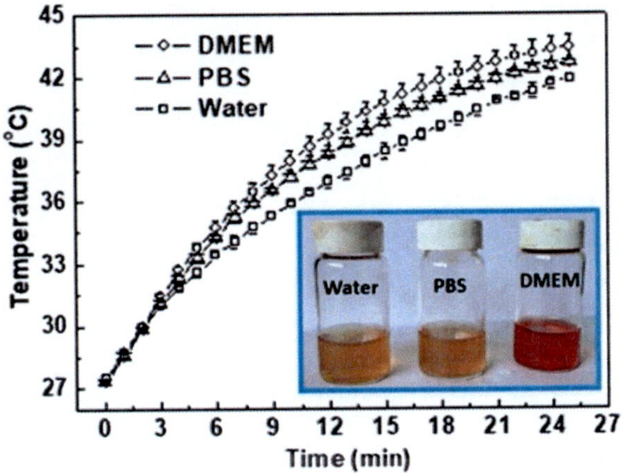

Fig. 7.1 Fe_3O_4 nanoparticles coated with proteins at a concentration of 1 mg/ml of Fe in different media (Reprinted from [22] with permission Elsevier)

7.1.2 Magnetic Separation

In the biomedical field, it is often necessary to separate specific biological elements (for example, DNA, proteins, and cells) from their natural environment. Superparamagnetic colloids of spinel ferrites are ideal for this application due to their nature of magnetization with and without an external magnetic field, allowing the transport of biomaterials within a magnetic field. In a typical separation procedure, biological entities are labeled by superparamagnetic colloids and then subjected to separation by an external magnetic field [23].

Thus, magnetic separation techniques are combined with spinel ferrite nanoparticles that bind to the elements of interest. Magnetic particles with a size of a few nanometers, such as superparamagnetic iron oxide particles, have been used extensively for the separation and purification of cells and biomolecules in bioprocesses.

Due to their small size and high surface area, spinel ferrites nanoparticles present many superior characteristics for applications in bioseparation compared to conventional micrometric size methods, such as resins or granules. Ferrite nanoparticles have an excellent dispersion capacity, fast and effective binding of biomolecules, and reversible and controllable flocculation [23].

7.1.3 Drug Delivery

Controlled drug delivery, one of the first areas to be developed in nanomedicine, arises with using the physical–chemical properties of nanomaterials to guide drugs to specific points in the body.

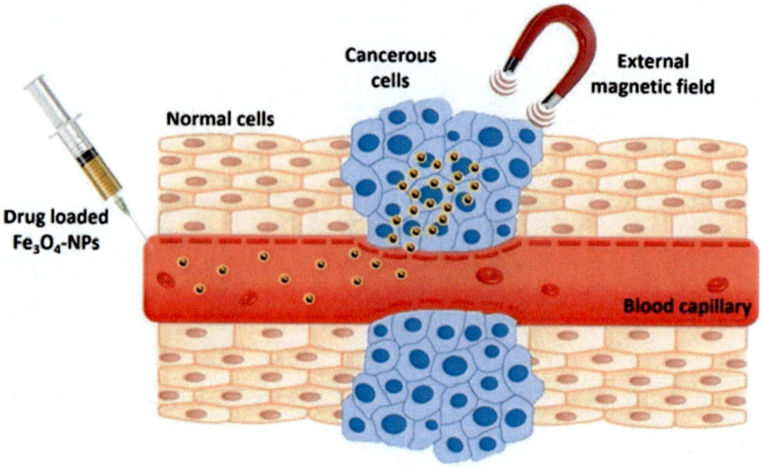

Fig. 7.2 Targeted drug delivery system using drug-loaded spinel ferrite nanoparticles (Reprinted from [24] with permission from Elsevier)

Drug segmentation has emerged as one of the most modern drug delivery technologies. The possibilities for applying magnetic spinel ferrite nanoparticles as a drug carrier have been increasing dramatically in recent years.

Magnetic spinel ferrite nanoparticles combined with an external magnetic field and/or magnetizable implants allow the particles to be delivered to the desired target area while the medication is released in fractional form (magnetic drug segmentation). Transporting drugs to a specific site eliminates side effects and even reduces the required dosage. As with magnetic hyperthermia, the surfaces of these particles are generally modified with organic polymers and inorganic metals or oxides to make them biocompatible and suitable for further functionalization by the execution of several bioactive molecules.

In addition, the temporalized delivery, the challenges of this line of research are the production of theranostic nanomaterials. These types of nanomaterials may switch between treatment and diagnosis capacity.

A targeted drug delivery system is illustrated in Fig. 7.2. The drug-loaded ferrite spinel is inserted in the human body by parenteral drug administration. The external magnetic field helps accumulating and releasing the drug at the desired site [24].

7.1.4 Magnetic Resonance Imaging (MRI)

Magnetic Resonance Imaging (MRI) is the primary medical technique for obtaining soft tissue images. Drug administration to patients is necessary to increase the contrast between normal and abnormal tissue and/or indicate the status of organic functions or blood flow. The technique is based on the alteration of the magnetization presented by the tissues from the period of relaxation of the protons present in the water molecule (hydrogen nuclei) under the action of a static solid field after the removal of another transverse magnetic field [18].

Magnetic nanoparticles, especially those smaller than 10 nm, represent an alternative class of contrast agents for nuclear magnetic resonance with physical advantages, as they further highlight the behavior of protons from different tissues. At the borderline between nanomaterials and medical diagnostics, superparamagnetic iron oxide nanoparticles prove to be a class of new probes useful for in vitro and in vivo procedures in cellular and molecular imaging.

The face-centered cubic structure of spinel ferrites allows electrons to jump between iron ions that occupy the tetrahedral and octahedral interstitial sites, thus giving them the properties of the semi-metallic molecules suitable for magnetic resonance imaging [25].

Superparamagnetic contrast agents have the advantage of producing enhanced proton relaxation in MRI when compared to paramagnetic nanoparticles. Consequently, lesser amounts of a superparamagnetic iron oxide nanoparticle agent are needed for imaging. For their application as magnetic fluids in MRI, superparamagnetic iron oxide nanoparticles should be dispersed in a biocompatible and biodegradable carrier.

The enhancement of the contrasts used in MRI has become essential for imaging the human body. Biological tissue contains a high concentration of protons in the form of water and fat [25]. Superparamagnetic spinel ferrites do not present any hysteresis above the blocking temperature, so they can be used as MRI augmentation agents that do not maintain any magnetization when the magnetic is removed.

7.2 Electrical and Electronic Devices

Currently, the greatest portion of the world's demand for ferrites is directed at the production of electrical and electronic devices. These materials are commonplace in several fields, from the large (inductors and transformers) to the tiniest equipment—as sensors and filters in circuit boards, for instance. A few of these numerous applications will be presented in this chapter.

7.2.1 EMI Filters

Electromagnetic interference (EMI) is a form of disturbance to an electric circuit caused by the passage of unwanted currents. These events may be caused by the device itself and from external sources. The occurrence of EMI may disrupt the proper functioning of the device or may cause unwanted events to occur. Given the sensitivity of several modern appliances, this noise may cause malfunctions or damage to the equipment. Sources of EMI may be natural, in the form of lightning strikes or solar flares, for instance, or artificial, such as microwave ovens and toasters. Whatever their source, these disruptions may result in severe problems, such as damaging communication networks or crucial medical equipment. Given its consequences, EMI suppression in electronic equipment is usually legally required in most countries around the world.

An effective way to protect a circuit from the impacts of electromagnetic interference is the utilization of EMI filters. When present in a circuit, these filters suppress the noise conducted in the system. Since the majority of EMI is in the high-frequency range, typical EMI filters are low-pass filters; in laymen's terms, these gadgets remove the high-frequency noise while letting the low-frequency signal through. Interference may be blocked either from exiting or entering the system in which the filter is placed.

The most common materials utilized in EMI filters are spinel ferrites. More specifically, NiZn and MnZn ferrites make up the majority of the commercially available beads. These devices can be found in the majority of the typical electronic devices one encounters in their daily life. A typical example of the utilization of ferrite beads is shown in Fig. 7.3, which depicts their application as a filter on a computer power cable.

The primary mechanism by which ferrite beads filter out EMI is via heat dissipation. The 50–60 Hz frequencies from the mains suffer little interference from the presence of the ferrite; signals at these frequencies should pass the device mostly unimpeded. At the higher frequencies associated with the most disruptive EMI (MHz to GHz), the impedance generated by the bead sifts these high signals out of the system. The high loss factor associated with NiZn and MnZn results in the dissipation of high frequencies into resistive heat. The generated heat is practically negligible in most settings. Despite their high frequencies, the energy transported by EMIs is usually quite small.

Both material and geometry must be considered when choosing the appropriate ferrite bead for an application. NiZn ferrite is usually the material utilized for EMI suppression at the megahertz range, where most of the disruptive noise is found.

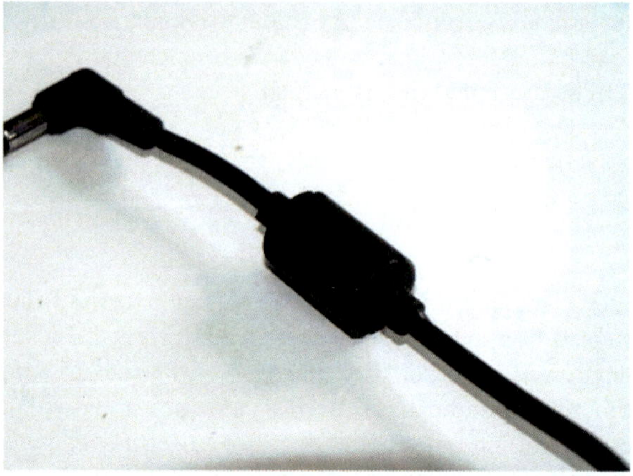

Fig. 7.3 Ferrite bead on a computer power cable (*Source* "Ferrite bead" by Santeri Viinamäki licensed under CC BY-SA 4.0)

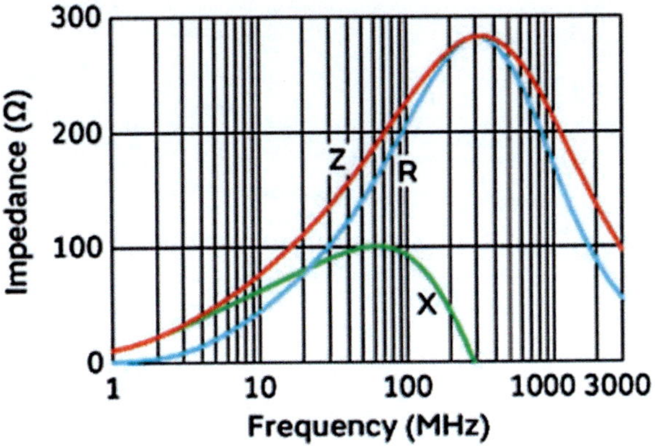

Fig. 7.4 Impedance curves of the ferrite bead BLM18PG221SN1, as provided by the manufacturer

MnZn is most effective at kHz frequencies, with some efficiency for the low MHz range. An example of the AC behavior of a commercial ferrite bead is shown in Fig. 7.4.

7.2.2 *Sensors*

As an example of the most recent electronic applications, ferrites are currently being considered for application as active material in several sensor arrays. Their electrical and magnetic properties have been successfully applied in the detection of chemical and environmental factors.

Stress sensors based on spinel ferrites take advantage of their magnetoelastic coupling. When a magnetic field is applied to ferromagnetic materials, their components rotate to align their easy axis to the applied field. This elongation/compression phenomenon generates a mechanical strain in the material. The reverse effect is observed when a mechanical stress is applied to a magnet. The reorganization of the structure affects the magnetic susceptibility of the material. This phenomenon is known as the Villari effect or inverse magnetostrictive effect.

Stress sensors based on the magnetostrictive effects measure the changes in magnetic properties remotely. The gathered data is subsequently converted into a mechanical stress value that would generate the measured variation in magnetic character. Cobalt ferrite shows great promise for application as stress sensors due to its large magnetostrictive effect and high sensitivity. Nevertheless, their magnetomechanical hysteresis is a drawback in the utilization of this material.

Brito and collaborators [26] have proposed the substitution of most of the cobalt atoms in $CoFe_2O_4$ by nickel. The authors prepared (Ni, Co)Fe_2O_4 samples via solid-state sintering and liquid-phase sintering. Both methods led to the formation of similar compositions, with 80–85% nickel as the divalent cation. The samples presented magnetoelastic sensitivities close to that of cobalt ferrite. Furthermore, the

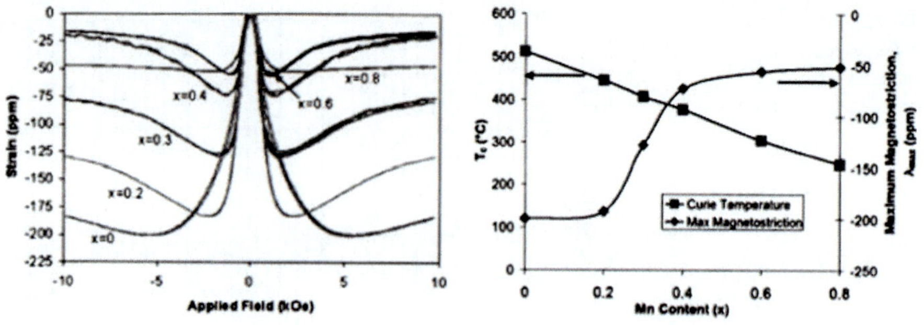

Fig. 7.5 Magnetostriction curves of $CoMn_xFe_{2-x}O_4$, and their maximum magnetostriction and Curie temperature as a function of manganese content (Adapted from [27] with permission from AIP Publishing)

hysteresis was significantly less pronounced, which would enable the application of these materials as mechanical stress sensors.

The substation of iron in spinel ferrites has also shown great promise for the improvement of $CoFe_2O_4$ sensors. Paulsen et al. [27] have studied the magnetostrictive behavior of $Co(Fe,Mn)_2O_4$ materials, aiming to reduce the hysteresis of these sensors at room temperature. Since the magnetic properties are influenced by the Curie temperature of the compound, the authors attempted to affect this critical point without great loss of magnetic character. The magnetostriction curves of the Mn-substituted ferrites are shown in Fig. 7.5. Although the intensity of the effect is decreased with higher manganese contents, the slopes at low fields are not adversely affected. The Curie temperatures decreased linearly with increasing substitution, showing that the thermal dependence was also affected, which should in turn affect the hysteresis at lower temperatures.

Spinel ferrite sensors may also be employed in the detection of chemical species. The interactions between the sensor and the compound of interest are varied and differ for each and every detection system. For the sake of brevity, a few selected examples of H_2 sensing systems will be presented.

Hydrogen gas is a flammable and explosive gas present in several industrial operations. Due to the hazards associated with its use, the detection of even small amounts of this gas in the environment is of vital importance, as H_2 leakage may lead to serious accidents. Mukherjee and Majumder [28] have studied the response of porous nanocrystalline $ZnFe_2O_4$ as a hydrogen gas sensor. The resistivity of zinc ferrite is decreased in the presence of H_2, and this decrease is proportional to the amount of gas present in the environment (Fig. 7.6). The authors proposed a specific mechanism for the change in the conductivity of this n-type semiconductor in the presence of adsorbed hydrogen. The sensitivity of the sensor at a fixed concentration of 1600 ppm of H_2 is also shown in Fig. 7.6.

Falsafi and collaborators [29] have studied the effect of samarium-doping on the H_2 sensing capability of $CoFe_2O_4$. The authors observed that the presence of the rare-earth increased the electrical response of the sensor compared to pure cobalt

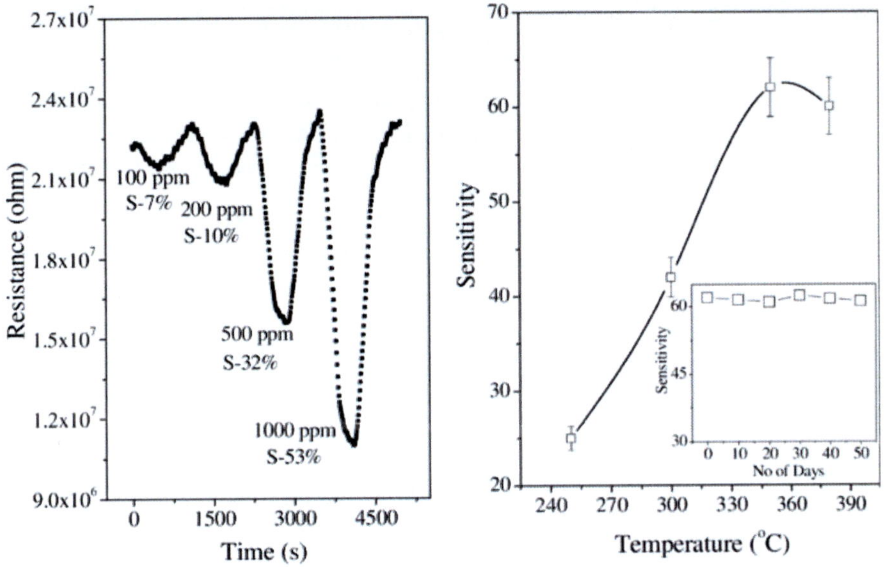

Fig. 7.6 **Left** Resistance of $ZnFe_2O_4$ in the presence of different concentrations of H_2. **Right** Sensitivity of this material at different temperatures in the presence of 1600 ppm H_2. The inset shows the temporal stability of the sensor after a few weeks (Adapted from [28] with permission from AIP Publishing)

ferrite. An interesting aspect considered in this study was the effect of interfering gases, as an ideal sensor should be highly selective toward a specific analyte. The authors reported the responses to CO, NO_2, CH_4, ethanol, and CO_2, which are shown in Fig. 7.7. The authors note that the response of the spinel ferrite to other common gases is smaller than that of the alarm threshold for H_2.

Fig. 7.7 Responses of a Sm-$CoFe_2O_4$ sensor to different gases (Reprinted from [29] with permission from Elsevier)

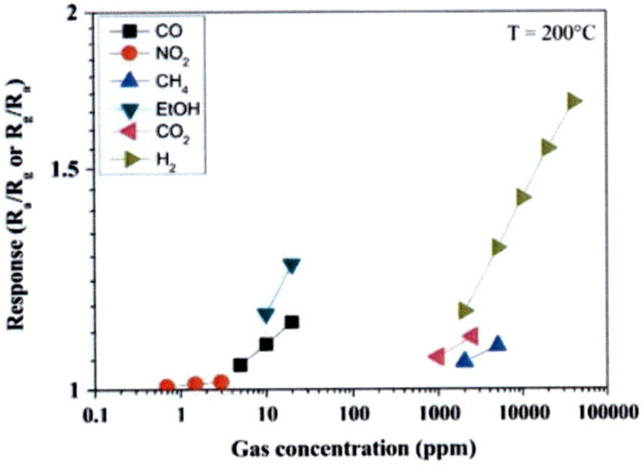

7.3 Adsorption

One of the most exciting applications of spinel ferrites is their use as adsorbents in the purification of effluents or contaminated water bodies. In general, most contaminants can be removed from water bodies via flocculation/coagulation processes. Nevertheless, emerging organic contaminants and harmful inorganic ions can prove very hard to remove via the strategies commonly utilized in water treatment facilities. In the presence of such pollutants, a final polishing of the effluent is usually necessary. Adsorption is one of the said processes, and it is commonly utilized worldwide.

One of the decisive factors when choosing an adsorbent is a high surface area, which would increase and accelerate the removal of contaminants. In this regard, nanoparticles would be ideally suited for application as adsorbents, owing to their inherent high surface area. Nevertheless, despite their high adsorption capacity, nanoparticles pose another problem: their separation from the effluent is also challenging. The application of run-of-the-mill nanoparticles requires the utilization of either a support system that would impede their transfer to the fluid or of a costly secondary separation process.

Ferrite nanoparticles can easily circumvent this problem. Their inherent magnetism leads to a straightforward separation process; a regular electromagnet ensures that no adsorbent is lost in the process. This phenomenon can be seen in Fig. 7.8. This factor, coupled with environmental and health considerations, enables the utilization of ferrite nanoparticles as a cost-effective advanced adsorbent in the purification of contaminated water bodies. Indeed, these materials are currently applied in the recovery of oil from oil spills in the ocean. While some of the adsorption capabilities of ferrites have already been discussed in previous chapters, this section will further discuss the application of these materials in model and existing systems.

Several metal ions are of particular interest due to their harmful interactions with the human body and the environment. Metallic ions are linked to several diseases and

Fig. 7.8 Disperse $ZnFe_2O_4$ nanoparticles and their separation from solution via the application of a magnetic field (Adapted from [30] with permission from Elsevier)

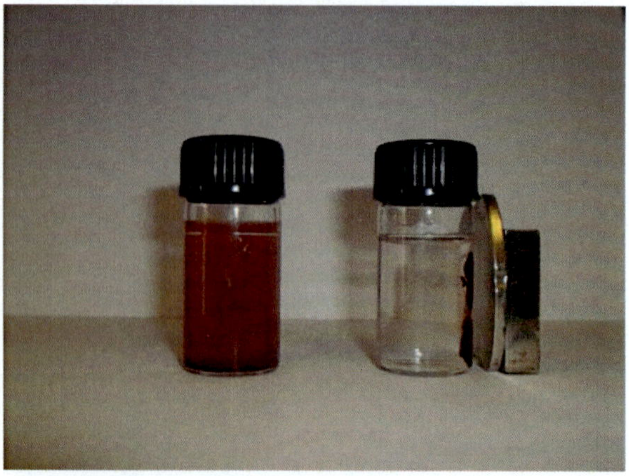

environmental disasters, and their removal from effluents can be tricky for a variety of reasons. The adsorption of these ions onto ferrite nanoparticles shows great promise. Jung and coworkers [31] have demonstrated the capability of manganese ferrite in the adsorption of heavy metals from water. The authors show that a $MnFe_2O_4$/biochar composite may be used to remove Pb^{2+}, Cd^{2+}, and Cu^{2+} ions from water. The material shows greater selectivity toward the removal of lead from the solution. Thermodynamic data shows that DG° becomes more negative with increasing temperature, indicating a higher spontaneity likely linked to the higher desolvation rates of heavy metals. Positive enthalpy values show that the adsorption process is endothermic, which insinuates that the endothermicity of desolvation exceeds the exothermal character of the other steps.

Complex inorganic ions may also be recovered from solution via adsorption onto ferrites. Zirconium ferrite has shown promise in the removal of phosphates from sewage, thus helping prevent eutrophication in the streams where the effluents are discharged [32]. The systems proposed by Ito and collaborators managed to remove 90% of the phosphate of a solution containing 0.86 mg/L of this pollutant in only five minutes. Filtration velocities of 1 m/s and the regeneration properties of the adsorbent are also excellent for application in large-scale settings.

A particular case of inorganic contaminants is shown in the literature to be readily adsorbed onto ferrites: radionuclides. These radioactive elements are—for obvious reasons—extremely hazardous to the environment; therefore, their effective removal from water bodies is crucial in decontamination processes. Tan and collaborators [33] have recently shown the effectiveness of a $CoFe_2O_4$/carbon nanotube composite in removing U^{6+} from water. A maximum in adsorption capacity of 212.7 mg/g was achieved at 25 °C. The capacity of the composite was higher than that of the separate constituents, though all three systems reached their highest capacity at a similar pH range (5–6). Reusability studies show 90.2% desorption of U(VI) after treatment in a 0.5M $NaHCO_3$ solution. After three adsorption–desorption cycles, the adsorption capacity decreases less than 10%. Similar systems based on $MgFe_2O_4$ composites show even greater adsorption capacities, of over 2200 mg/g [34].

Thorium is a toxic radionuclide with great promise in advanced power generation. Thus, its optimized recovery is both an environmental and an economic issue. Alarqabeh and collaborators [35] have assessed the performance of $MnFe_2O_4$ nanoparticles in the recovery of U^{6+} and Th^{4+} ions. The authors report that the adsorption of thorium is explained by a pseudo-second-order equilibrium model. A pH of 3 was defined as the best condition for the efficient removal of this radionuclide, with close to 90% of metal uptake. This phenomenon was shown to occur via an exothermic process.

Some selected adsorbate/adsorbent systems are shown in Table 7.1, which also exhibits the removal capacity of the ferrites and the analyzed conditions.

Ferrites have also shown great potential in the removal of emergent organic contaminants (EOC) in a similar manner to the systems presented previously for the recovery of inorganic compounds. Nonetheless, a further factor may increase the performance of ferrites in the decontamination of water polluted with EOCs: the possibility of decomposition of these compounds via Fenton reactions.

Table 7.1 Examples of pollutant/ferrite adsorption systems

Adsorbent	Size (nm)	Specific area (m^2/g)	Pollutant	Capacity (mg/g)	Reference
$CuFe_2O_4$/DC	17.91	16.96	Pb(II)	921.1	[36]
$MnFe_2O_4$/GO	20	67.5	As(V)	240.4	[37]
$MnFe_2O_4$	20	81.09	As(III)	49.5	[38]
$CoFe_2O_4$	20	–	Cs(I)	75	[39]
Fe_3O_4/CFA	–	–	U(VI)	328.4	[40]
$NiFe_2O_4$	41	122.7	Th(IV)	88.49	[41]

As shown in previous chapters, ferrites may be used as heterogeneous Fenton catalysts. In its original form, Fenton technology is commonplace in the final polishing of wastewater. However, this process generates an iron sludge which incurs a greater cost for the treatment plant. The utilization of ferrites as heterogeneous catalysts overcomes this hurdle. The utilization of magnetic solids as Fenton catalysts facilitates the separation of the catalyst from the reactional medium. Furthermore, the reaction would primarily decompose the materials adsorbed onto the surface of the solids. The synergy between adsorption and decomposition further improves the performance of these spinels in the decontamination of water bodies.

Anchieta et al. [42] assessed the Fenton performance of a zinc ferrite prepared via a microwave-assisted method in the degradation of Procion red. $ZnFe_2O_4$ treated at 1000 W for 30 min showed the fastest decolorization rate (0.086 min^{-1}) for the studied dye. The authors have also assessed an important property that is commonly sidelined in dye degradation studies: the presence of colorless products. Instead of directly achieving full mineralization, dyes may degrade toward complex molecules that are not visible via the applied spectrophotometric methods. The analysis of TOC (total organic carbon) showed that the mineralization rates were actually quite slower than the apparent rate, on the order of 0.003 min^{-1}. The results are a great reminder that mineralization usually takes longer than it may seem based only on visual clues.

Another class of persistent organic pollutants is chlorophenols. These compounds are mainly released via plastics and pesticide industries, and they display high toxicity and low biodegradability. Therefore, their safe removal from streams is a great concern. Kurian and Nair [43] have shown that spinel ferrites are effective Fenton catalysts in the complete mineralization of chlorophenols. The authors have studied materials with the composition $Ni_xZn_{1-x}Fe_2O_4$ with $x = 0.0, 0.25, 0.5, 0.75$, and 1.0. The results of the catalytic degradation of 4-chlorophenol show a 99.69% reduction in the chemical oxygen demand (COD) after 105 min of reaction. The catalytic activity increased with Zn substitution, with the terminal member $ZnFe_2O_4$ almost completely removing the pollutant in 45 min. The catalysts were reusable for five consecutive cycles with minor loss of activity and negligible iron leaching.

Wang et al. [44] assessed the Fenton activity of several ferrites on the degradation of imidacloprid, an important pesticide. Nanosized Fe_3O_4, conventional $CuFe_2O_4$ and $CoFe_2O_4$, and mesoporous $CuFe_2O_4$ and $CoFe_2O_4$ were compared. The apparent degradation rate for mesostructured copper ferrite was 1.0445 h^{-1}, the highest among

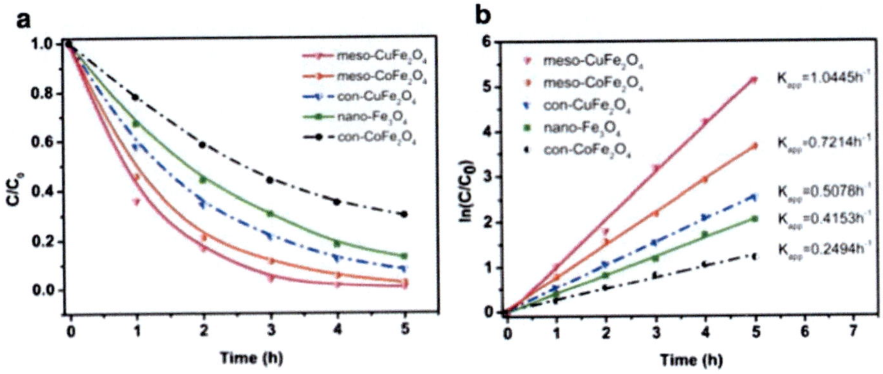

Fig. 7.9 Removal efficiency of imidacloprid by different spinel ferrite systems. Conditions: pH 3.00, 40 mM H_2O_2, 0.3 g.L^{-1} catalyst dosage, and 10 mg.L^{-1} imidacloprid (**A**))(Adapted from [44] with permission from Elsevier)

all analyzed ferrites, as shown in Fig. 7.9. The authors assign this improved catalytic activity of $CuFe_2O_4$ to the redox recycle of Fe^{2+}/Fe^{3+} and Cu^{+}/Cu^{2+}, with the thermodynamically favored reduction of Fe^{3+} by Cu^{+}. Magnetic mesoporous copper ferrite has shown to be stable, reusable, and easy to separate from the reactional medium.

Ammonia may also be fully mineralized via heterogeneous Fenton technology. A graphene/manganese ferrite composite was used by Zhou and collaborators to degrade this recalcitrant water pollutant [45]. The authors report a 92% degradation of ammonia to N_2 under visible light. The photo-Fenton system followed a pseudo-first-order behavior, with an apparent rate of 0.0369 min^{-1}. XPS analysis indicates the presence of Mn^{3+} and Fe^{2+}, suggesting that both metals actively participate in the photocatalytic redox reaction.

References

1. Kefeni KK, Msagati TAM, Mamba BB (2017) Ferrite nanoparticles: synthesis, characterisation and applications in electronic device. Mater. Sci. Eng. B Solid-State Mater. Adv. Technol 215:37–55
2. Chatterjee DK, Diagaradjane P, Krishnan S (2011) Nanoparticle-mediated hyperthermia in cancer therapy. Ther Deliv 2:1001–1014. https://doi.org/10.4155/tde.11.72
3. Ito A, Shinkai M, Honda H, Kobayashi T (2005) Medical application of functionalized magnetic nanoparticles. J Biosci Bioeng 100:1–11. https://doi.org/10.1263/JBB.100.1
4. Nuzhina J V., Shtil AA, Prilepskii AY, Vinogradov VV (2019) Preclinical evaluation and clinical translation of magnetite-based nanomedicines. J. Drug Deliv. Sci. Technol 54:101282
5. Amiri M, Salavati-Niasari M, Akbari A (2019) Magnetic nanocarriers: evolution of spinel ferrites for medical applications. Adv Colloid Interface Sci 265:29–44
6. Bañobre-López M, Teijeiro A, Rivas J (2013) Magnetic nanoparticle-based hyperthermia for cancer treatment. Reports Pract. Oncol. Radiother 18:397–400

7. Das P, Colombo M, Prosperi D (2019) Recent advances in magnetic fluid hyperthermia for cancer therapy. Colloids Surfaces B Biointerfaces 174:42–55. https://doi.org/10.1016/J.COLSURFB.2018.10.051
8. Lévy M, Wilhelm C, Siaugue JM et al (2008) Magnetically induced hyperthermia: size-dependent heating power of γ-Fe_2O_3 nanoparticles. J Phys Condens Matter. https://doi.org/10.1088/0953-8984/20/20/204133
9. Kafrouni L, Savadogo O (2016) Recent progress on magnetic nanoparticles for magnetic hyperthermia. Prog Biomater. https://doi.org/10.1007/s40204-016-0054-6
10. Espinosa A, Di Corato R, Kolosnjaj-Tabi J et al (2016) Duality of iron oxide nanoparticles in cancer therapy: amplification of heating efficiency by magnetic hyperthermia and photothermal bimodal treatment. ACS Nano. https://doi.org/10.1021/acsnano.5b07249
11. Hedayatnasab Z, Abnisa F, Daud WMAW (2017) Review on magnetic nanoparticles for magnetic nanofluid hyperthermia application. Mater Des. https://doi.org/10.1016/j.matdes.2017.03.036
12. Hilger I (2013) In vivo applications of magnetic nanoparticle hyperthermia. Int J Hyperth 29:828–834
13. Johannsen M, Gneveckow U, Eckelt L et al (2005) Clinical hyperthermia of prostate cancer using magnetic nanoparticles: presentation of a new interstitial technique. Int J Hyperth. https://doi.org/10.1080/02656730500158360
14. Di Corato R, Béalle G, Kolosnjaj-Tabi J et al (2015) Combining magnetic hyperthermia and photodynamic therapy for tumor ablation with photoresponsive magnetic liposomes. ACS Nano. https://doi.org/10.1021/nn506949t
15. Hou CH, Hou SM, Hsueh YS et al (2009) The in vivo performance of biomagnetic hydroxyapatite nanoparticles in cancer hyperthermia therapy. Biomaterials. https://doi.org/10.1016/j.biomaterials.2009.04.020
16. Du L, Zhou J, Wang X et al (2009) Effect of local hyperthermia induced by nanometer magnetic fluid on the rabbit VX2 liver tumor model. Prog Nat Sci. https://doi.org/10.1016/j.pnsc.2009.06.006
17. Mallory M, Gogineni E, Jones GC et al (2016) Therapeutic hyperthermia: the old, the new, and the upcoming. Crit Rev Oncol Hematol 97:56–64
18. Berry CC, Curtis ASG (2003) Functionalisation of magnetic nanoparticles for applications in biomedicine. J Phys D Appl Phys 36:R198–R206. https://doi.org/10.1088/0022-3727/36/13/203
19. Dadfar SM, Roemhild K, Drude NI et al (2019) Iron oxide nanoparticles: diagnostic, therapeutic and theranostic applications. Adv Drug Deliv Rev 138:302–325
20. Venturini J, Tonelli AM, Wermuth TB et al (2019) Excess of cations in the sol-gel synthesis of cobalt ferrite ($CoFe_2O_4$): a pathway to switching the inversion degree of spinels. J Magn Magn Mater 482:1–8. https://doi.org/10.1016/j.jmmm.2019.03.057
21. Tonelli AM, Venturini J, Arcaro S et al (2020) Novel core-shell nanocomposites based on TiO_2-covered magnetic Co_3O_4 for biomedical applications. J Biomed Mater Res Part B Appl Biomater 108:1879–1887. https://doi.org/10.1002/jbm.b.34529
22. Gawali SL, Shelar SB, Gupta J et al (2021) Immobilization of protein on Fe_3O_4 nanoparticles for magnetic hyperthermia application. Int J Biol Macromol 166:851–860. https://doi.org/10.1016/j.ijbiomac.2020.10.241
23. Faraji M, Yamini Y, Rezaee M (2010) Magnetic nanoparticles: synthesis, stabilization, functionalization, characterization, and applications. J Iran Chem Soc 7:1–37
24. Yew YP, Shameli K, Miyake M et al (2018) Green biosynthesis of superparamagnetic magnetite Fe_3O_4 nanoparticles and biomedical applications in targeted anticancer drug delivery system: a review. Arab J Chem. https://doi.org/10.1016/J.ARABJC.2018.04.013
25. Chaughule RS, Purushotham S, Ramanujan KV (2012) Magnetic nanoparticles as contrast agents for magnetic resonance imaging. Proc Natl Acad Sci India Sect A - Phys Sci 82. https://doi.org/10.1007/s40010-012-0038-4
26. Brito VLO, Cunha SA, Araújo FF et al (2015) Processing and characterization of a Ni-Co ferrite for sensor applications. Cerâmica 61:341–349

27. Paulsen JA, Ring AP, Lo CCH et al (2005) Manganese-substituted cobalt ferrite magnetostrictive materials for magnetic stress sensor applications. J Appl Phys 97:44502. https://doi.org/10.1063/1.1839633
28. Mukherjee K, Majumder SB (2009) Analyses of response and recovery kinetics of zinc ferrite as hydrogen gas sensor. J Appl Phys 106:64912. https://doi.org/10.1063/1.3225996
29. Falsafi F, Hashemi B, Mirzaei A et al (2017) Sm-doped cobalt ferrite nanoparticles: a novel sensing material for conductometric hydrogen leak sensor. Ceram Int 43:1029–1037. https://doi.org/10.1016/j.ceramint.2016.10.035
30. Konicki W, Sibera D, Mijowska E et al (2013) Equilibrium and kinetic studies on acid dye Acid Red 88 adsorption by magnetic $ZnFe_2O_4$ spinel ferrite nanoparticles. J Colloid Interface Sci 398:152–160. https://doi.org/10.1016/j.jcis.2013.02.021
31. Jung K-W, Lee SY, Lee YJ (2018) Facile one-pot hydrothermal synthesis of cubic spinel-type manganese ferrite/biochar composites for environmental remediation of heavy metals from aqueous solutions. Bioresour Technol 261:1–9. https://doi.org/10.1016/j.biortech.2018.04.003
32. Ito D, Nishimura K, Miura O (2009) Removal and recycle of phosphate from treated water of sewage plants with zirconium ferrite adsorbent by high gradient magnetic separation. J Phys Conf Ser 156:012033. https://doi.org/10.1088/1742-6596/156/1/012033
33. Tan L, Liu Q, Jing X et al (2015) Removal of uranium(VI) ions from aqueous solution by magnetic cobalt ferrite/multiwalled carbon nanotubes composites. Chem Eng J 273:307–315. https://doi.org/10.1016/j.cej.2015.01.110
34. Li X, Li Y, Wu Q et al (2019) Ultrahigh uranium uptake by magnetic magnesium ferrite loaded hydrothermal carbon nanosheets under acidic condition. Chem Eng J 365:70–79. https://doi.org/10.1016/j.cej.2019.02.002
35. Alaqarbeh M, Khalili FI, Kanoun O (2020) Manganese ferrite ($MnFe_2O_4$) as potential nanosorbent for adsorption of uranium(VI) and thorium(IV). J Radioanal Nucl Chem 323:515–537. https://doi.org/10.1007/s10967-019-06953-4
36. Khan MA, Otero M, Kazi M et al (2019) Unary and binary adsorption studies of lead and malachite green onto a nanomagnetic copper ferrite/drumstick pod biomass composite. J Hazard Mater 365:759–770. https://doi.org/10.1016/J.JHAZMAT.2018.11.072
37. Lan Huong PT, Tu N, Lan H et al (2018) Functional manganese ferrite/graphene oxide nanocomposites: effects of graphene oxide on the adsorption mechanisms of organic MB dye and inorganic As(v) ions from aqueous solution. RSC Adv 8:12376–12389. https://doi.org/10.1039/C8RA00270C
38. Martinez–Vargas S, Martínez AI, Hernández–Beteta EE et al (2018) As(III) and As(V) adsorption on manganese ferrite nanoparticles. J Mol Struct 1154:524–534. https://doi.org/10.1016/j.molstruc.2017.10.076
39. Hassan MR, Aly MI (2019) Adsorptive removal of cesium ions from aqueous solutions using synthesized Prussian blue/magnetic cobalt ferrite nanoparticles. Part Sci Technol 1–11. https://doi.org/10.1080/02726351.2018.1539799
40. Chen Z, Wang J, Pu Z et al (2017) Synthesis of magnetic Fe_3O_4/CFA composites for the efficient removal of U(VI) from wastewater. Chem Eng J 320:448–457. https://doi.org/10.1016/j.cej.2017.03.074
41. Lingamdinne LP, Choi Y-L, Kim I-S et al (2017) Preparation and characterization of porous reduced graphene oxide based inverse spinel nickel ferrite nanocomposite for adsorption removal of radionuclides. J Hazard Mater 326:145–156. https://doi.org/10.1016/J.JHAZMAT.2016.12.035
42. Anchieta CG, Severo EC, Rigo C et al (2015) Rapid and facile preparation of zinc ferrite ($ZnFe_2O_4$) oxide by microwave-solvothermal technique and its catalytic activity in heterogeneous photo-Fenton reaction. Mater Chem Phys 160:141–147. https://doi.org/10.1016/j.matchemphys.2015.04.016
43. Kurian M, Nair DS (2015) Heterogeneous Fenton behavior of nano nickel zinc ferrite catalysts in the degradation of 4-chlorophenol from water under neutral conditions. J Water Process Eng 8:e37–e49. https://doi.org/10.1016/j.jwpe.2014.10.011

44. Wang Y, Zhao H, Li M et al (2014) Magnetic ordered mesoporous copper ferrite as a heterogeneous Fenton catalyst for the degradation of imidacloprid. Appl Catal B Environ 147:534–545. https://doi.org/10.1016/j.apcatb.2013.09.017
45. Zhou Y, Xiao B, Liu S-Q et al (2016) Photo-Fenton degradation of ammonia via a manganese–iron double-active component catalyst of graphene–manganese ferrite under visible light. Chem Eng J 283:266–275. https://doi.org/10.1016/j.cej.2015.07.049